Photovoltaic Systems and the National Electric Code

Used throughout the United States and many other countries, the National Electric Code (NEC) is the world's most detailed set of electrical codes pertaining to photovoltaic (PV) systems.

Photovoltaic Systems and the National Electric Code presents a straightforward explanation of the NEC in everyday language. The new book is based on the 2017 NEC, which will be used widely until 2023, with most of the interpretations and material staying true long after. This book interprets the distinct differences between previous versions of the NEC and the 2017 NEC and clarifies how these Code changes relate specifically to photovoltaic installations.

Written by two of the leading authorities and educators in the field, this book will be a vital resource for solar professionals, as well as anyone preparing for a solar certification exam.

Bill Brooks is Principal Engineer at Brooks Engineering, Vacaville, USA.

Sean White is a Solar PV professor, trainer and contractor in the USA.

Photovoltaic Systems and the National Electric Code

Bill Brooks and Sean White

First published 2018
by Routledge
2 Park Square, Milton Park, Abingdon, Oxon OX14 4RN

and by Routledge
711 Third Avenue, New York, NY 10017

Routledge is an imprint of the Taylor & Francis Group, an informa business

© 2018 Bill Brooks and Sean White

The right of Bill Brooks and Sean White to be identified as authors of this work has been asserted by them in accordance with sections 77 and 78 of the Copyright, Designs and Patents Act 1988.

All rights reserved. No part of this book may be reprinted or reproduced or utilised in any form or by any electronic, mechanical, or other means, now known or hereafter invented, including photocopying and recording, or in any information storage or retrieval system, without permission in writing from the publishers.

Trademark notice: Product or corporate names may be trademarks or registered trademarks, and are used only for identification and explanation without intent to infringe.

British Library Cataloguing-in-Publication Data
A catalogue record for this book is available from the British Library

Library of Congress Cataloging-in-Publication Data
A catalog record for this book has been requested

ISBN: 978-1-138-08752-1 (hbk)
ISBN: 978-1-138-08753-8 (pbk)
ISBN: 978-1-315-11030-1 (ebk)

Typeset in Sabon
by Apex CoVantage, LLC

Contents

List of figures	vii
List of tables	ix
Introduction	1
1 Article 690 photovoltaic (PV) systems	4
2 Article 690 photovoltaic systems part II circuit requirements	26
3 Section 690.12 rapid shutdown	58
4 Article 690 part III disconnecting means	74
5 Article 690 part IV wiring methods	88
6 Article 690 part V grounding and bonding	103
7 Article 690 part VI to the end of 690	123
8 Article 691 large-scale photovoltaic (PV) electric power production facility	130
9 Article 705 interconnected electric power production sources	138
10 Storage articles	168

11	Chapters 1–4, Chapter 9 tables and Informative Annex C	177
12	PV wire sizing examples	189
	Index	199

Figures

0.1	1895 Niagara Falls power plant	2
1.1	1984 NEC (a much smaller Code book)	5
1.2	2014 NEC Figure 690.1(a) PV power source	8
1.3	2017 NEC PV Figure 690.1(a) PV power source	9
1.4	Interactive system [2017 NEC Fig 690.1(b)]	10
1.5	Ac module system [2017 NEC Fig 690.1(b)]	10
1.6	Dc coupled multimode system [2017 NEC Fig 690.1(b)]	11
1.7	Ac coupled multimode system [2017 NEC Fig 690.1(b)]	13
1.8	Stand-alone system [2017 NEC Fig 690.1(b)]	14
2.1	IV curve with different currents plotted showing maximum circuit current, which is used for sizing wires, above and beyond short circuit current	37
2.2	Partial datasheet from outback stand-alone inverter	39
2.3	Module interconnect for multiple parallel-connected module circuits	47
2.4	Two PV source circuits backfeeding a short on another PV source circuit	53
2.5	Fuses listed for PV	53
2.6	Dangerous dc arc-fault (do not try this at home)	57
3.1	AP system 4 module inverter	63
3.2	Rapid shutdown initiation switch	66
3.3	NEC Figure 690.56(C)(1)(a) reduced array shock hazard sign	68
3.4	NEC Figure 690.56(C)(1)(b) conductors leaving array level rapid shutdown sign	69
3.5	Buildings with more than one rapid shutdown type example	71
3.6	Rapid shutdown sign	72
4.1	PV system disconnect sign	76
4.2	Finger safe fuse holder	84

6.1	Fuse grounded PV array with one functional grounded conductor	105
6.2	Bipolar PV array	107
6.3	Non-isolated inverter showing ground fault pathway	108
6.4	2017 NEC ungrounded PV array AKA transformer-isolated inverter	110
6.5	Solidly grounded PV array	111
9.1	Feeder image showing where different parts of the Code apply to different parts of the feeder	146
9.2	705.12(B)(2)(1)(a) sufficient feeder ampacity	147
9.3	705.12(B)(2)(1)(b) Overcurrent device protecting feeder	149
9.4	Solar tap rules	150
9.5	25-foot tap rule	152
9.6	100% rule	153
9.7	705.12(B)(2)(3)(b) 120% rule	153
9.8	120% rule with multiple solar breakers acceptable	155
9.9	705.12(B)(2)(3)(c) sum rule	155
9.10	705.12(B)(2)(3)(d) center fed 120% rule	156
9.11	705.12(B)(3) marking label indicating multiple sources	157
9.12	Breakers over 1000V prices	159
12.1	Nicola Tesla demonstrates how to truly understand 3-phase in 1899	196

Tables

2.1	NEC Table 690.7(a) voltage correction factors for crystalline and multicrystalline silicon modules	30
5.1	Table 690.1(A) correction factors (ambient temperature correction factors for temperatures over 30°C)	91
5.2	Table 690.31(E) minimum PV wire strands for moving arrays	96
6.1	NEC Table 250.122 EGC based on OCPD	116

Introduction
Photovoltaic is on the cover of the 2017 NEC!

Photovoltaic (PV) is growing fast, and the PV material in the National Electric Code (NEC) is changing faster than anything the NEC has seen since the days of Thomas Edison and Nikola Tesla hashing it out over dc vs. ac. It appeared that Tesla was right when 2-phase ac power[1] was installed at Niagara Falls and that ac was the way of the future, but the future is always unpredictable and with PV, dc is making a comeback.

This book is designed to relay to the layperson working in the PV industry the NEC PV-related material and changes as simply as possible, but not simpler. We hope that professional engineers (PEs) and sunburnt solar installers alike will comprehend this easy writing style and be entertained just enough to not be bored learning about a Code that has been known to work better than melatonin on a redeye flight.

Since this book is about PV, rather than starting at the beginning of the NEC, we will start with the most relevant article of the NEC, which is **Article 690 Photovoltaic (PV) Systems**; we will then cover the new **Article 691 Large-Scale Photovoltaic (PV) Electric Power Production Facility** which modifies Article 690 for large PV systems and then dive into the interconnections of **Article 705 Interconnected Electric Power Production Sources** where we understand how PV and other power sources can connect to and feed other power sources, such as the utility grid. The next articles we will cover are the articles on energy storage, which are the old **Article 480 Storage Batteries** and the new and more relevant in 2017 **Article 706 Energy Storage Systems**. While we are on the subject of energy storage, we will cover the new **Article 710 Stand-Alone Systems** (which was formerly 690.10) and this will lead us to another new and renewable themed **Article 712 Dc Microgrids**. We will then go back to the beginning of the NEC and look at Chapters 1 through 4 of the NEC, which apply to all wiring systems, including PV. We will see that, in covering the new and

2 Introduction

renewable PV centric articles, we already covered the more important parts of Chapters 1 through 4 used for PV systems and all electric installations, such as Article 250 Grounding and Bonding and Article 310 for wire sizing. There will be many times, when we are covering material in Article 690, that we will go back and forth to other articles, since this is the way to properly use the NEC.

The NEC is updated every three years with a new Code cycle. This edition of *Photovoltaic Systems and the National Electric Code* reflects the 2017 NEC and will discuss earlier versions of the NEC. When the 2020 NEC comes out, this material will not be obsolete; in fact more than half the PV in the United States is installed in places that adopt the NEC three years after a Code is released. For instance, the state with half of the solar in the US is California, and in California, the 2017 NEC is adopted in 2020 and used until the 2020 NEC is adopted in 2023. It is also interesting to note that the proposals for changes to the NEC are crafted three years earlier, so the material in the 2017 NEC was proposed in 2014 and will be used on a regular basis by inspectors until nine years later. Since the equipment changes

Figure 0.1 1895 Niagara Falls power plant

Courtesy Wikimedia

https://en.wikipedia.org/wiki/Adams_Power_Plant_Transformer_House#/media/File:Westinghouse_Generators_at_Niagara_Falls.jpg

so fast in the PV industry, the Code writers intentionally leave parts of the Code open-ended to make way for new inventions that you may come up with, which will save lives and may make you rich.

The 2017 NEC proposals for Article 690 and for other solar-relevant parts of the Code were first proposed at meetings at NREL in Colorado in 2014 and put on a Word document by Bill Brooks. This Word document grew, and the proposals were refined with a lot of input. These future Codes were later proposed to the top dogs at the National Fire Protection Association by Ward Bower (inventor of the grid-tied inverter) and Bill Brooks of NEC Code Making Panel 4 in Hilton Head, North Carolina.

Now is the time to take out your 2017 NEC and follow along to understand PV and the NEC.

Note

1 The first power plant at Niagara Falls had two phases that were 90 degrees out of phase with each other (weird). Now we use three phases that are 120 degrees out of phase with each other. This is interesting!

1 Article 690 photovoltaic (PV) systems

Article 690 first came out in a little book known as the 1984 NEC and has been updated and mostly lengthened ever since.

In comparing the original 1984 version of Article 690 to today's NEC, there are many similarities yet also quite a few differences. Time to dig in!

Let us first list what we are dealing with in Article 690 before we dig deep. This will give us perspective and familiarize us with how to look things up quickly.

The NEC is also known as NFPA 70 and is divided into Chapters, then Articles, then Parts and Sections.

For example, rapid shutdown requirements are found in:

NEC Chapter 6 Special Equipment
Article 690 Solar Photovoltaic (PV) Systems
Part II Circuit Requirements
Section 690.12 Rapid Shutdown of PV Systems on Buildings

Here is what we find in Article 690:

Article 690 solar photovoltaic (PV) systems

Part I general (part)

690.1 Scope [Section 690.1]
690.2 Definitions [There are more NEC definitions in Article 100, such as the definitions for PV, ac and dc.]
690.4 General Requirements [They could not come up with a better title for this category.]
690.6 Alternating Current (ac) Modules

Article 690 photovoltaic (PV) systems 5

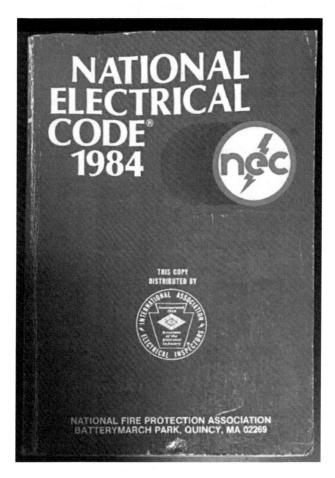

Figure 1.1 1984 NEC (a much smaller Code book)
Photo by Sean White

Part II circuit requirements

690.7 Maximum Voltage
690.8 Circuit Sizing and Current
690.9 Overcurrent Protection [Article 240 is also Overcurrent Protection.]
690.10 Stand Alone Systems [This has been moved to Article 710 in the 2017 NEC.]

6 *Article 690 photovoltaic (PV) systems*

690.11 Arc-Fault Circuit Protection (Direct Current)
690.12 Rapid Shutdown of PV Systems on Buildings [Big changes]

Part III disconnecting means

690.13 Photovoltaic System Disconnecting Means
690.15 Disconnection of PV Equipment

Part IV wiring methods

690.31 Methods Permitted
690.32 Component Interconnections
690.33 Connectors
690.34 Access to Boxes

Part V grounding and bonding [Article 250 is also grounding and bonding.]

690.41 System Grounding [Big changes in the 2017 NEC]
690.42 Point of System Grounding Connections
690.43 Equipment Grounding and Bonding
690.45 Size of Equipment Grounding Conductors
690.46 Array Equipment Grounding Conductors
690.47 Grounding Electrode System [Experts argue over a lot of this article, which is interesting to observe.]
690.50 Equipment Bonding Jumpers

Part VI marking

690.51 Modules
690.52 Alternating Current Photovoltaic Modules
690.53 Direct Current Photovoltaic Power Source
690.54 Interactive System Point of Interconnection
690.55 Photovoltaic Systems Connected to Energy Storage Systems
690.56 Identification of Power Sources [This includes new Rapid Shutdown signs.]

Part VII connection to other sources

690.59 Connection to Other Sources [Directs us to Article 705]

Part VIII energy storage systems

690.71 General [Directs us to Article 706]
690.72 Self- Regulated PV Charge Control

Now it is time to dive into the detail of Article 690.

Article 690 solar photovoltaic (PV) systems
Part I general (part)

690.1 scope (section 690.1)

Word-for-word NEC:

> "690.1 Scope. This article applies to solar PV systems, other than those covered by Article 691, including the array circuit(s), inverter(s), and controller(s) for such systems. [See Figure 690.1(a) and Figure 690.1(b).] The systems covered by this article may be interactive with other electrical power production sources or stand-alone or both, and may or may not be connected to energy storage systems such as batteries. These PV systems may have ac or dc output for utilization.
>
> Informational Note: Article 691 covers the installation of large-scale PV electric supply stations."

Discussion: For the most part 690.1 is self-explanatory, however, if we read the 2014 and the 2017 NEC carefully, we will notice that energy storage systems (batteries) are no longer part of the PV system.
2017 NEC language:

> "may or may not be connected to energy storage systems."

2014 NEC language:

> "may be interactive with other electrical power production sources or stand-alone, with or without electrical energy storage such as batteries."

It takes some careful analysis of the language, but we see that being connected to batteries in the 2017 NEC is different than with batteries in the 2014 NEC.

So what does this mean for us? Batteries are no longer part of the PV system as of the 2017 NEC and are part of a separate energy storage system that is covered in the new Article 706. Consequently, rapid shutdown and other requirements that are specific to PV systems no longer apply to the batteries.

8 *Article 690 photovoltaic (PV) systems*

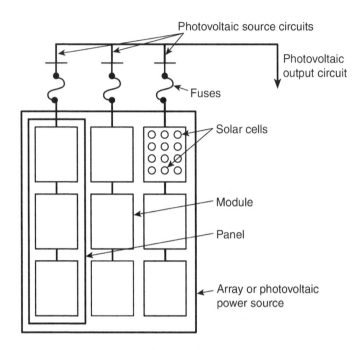

Figure 1.2 2014 NEC Figure 690.1(a) PV power source
Courtesy NFPA

Next, we see diagrams that will show us the dividing line between the PV system and *not* the PV system.

Section 690.1 also has some figures that we can look at in order to get a picture of what we are talking about.

Figure 1.2 is an image from the 2014 NEC.

Figure 1.3 is an image from the 2017 NEC.

Figure 1.3 is from the 2017 NEC with the added dc-to-dc converter.

From comparing these images, the main difference here is the insertion of the dc-to-dc converters. The writers of the NEC left the dc-to-dc converter definition open-ended for your billion-dollar invention. 2017 dc-to-dc converters are usually one per module, rather than three modules per converter in this image. Take note that, as we will learn coming up in Section 690.12 Rapid Shutdown, in 2019 the 2017 NEC will increase requirements for rapid shutdown on buildings and module level shutdown may be one of the only methods to comply. However, new inventions in the meantime could introduce other methods not currently foreseen.

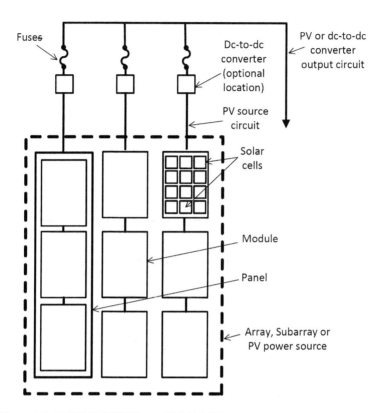

Figure 1.3 2017 NEC PV Figure 690.1(a) PV power source
Courtesy NFPA

> It is interesting to note that the solar cells in the diagram have gone from round in the 2014 NEC (really old style) to square in the 2017 NEC (polycrystalline). For someone first learning about solar, it could be confusing to see a solar module with 12 cells and then to see panels made of three modules. It would be even more confusing to have one dc-to-dc converter per three modules that is being connected with fuses to a dc-to-dc converter combining busbar and then off to a dc-to-dc converter output circuit. Dc-to-dc converters being installed in 2017 have a single PV module with a dc-to-dc converter under the module and then a number of dc-to-dc converters connected in series, and then the dc-to-dc converter source circuit is connected directly to the inverter.

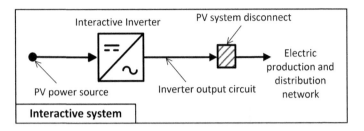

Figure 1.4 Interactive system [2017 NEC Fig 690.1(b)]
Courtesy NFPA

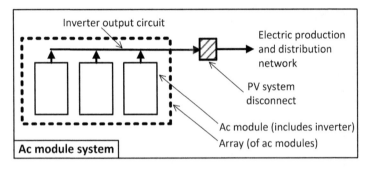

Figure 1.5 Ac module system [2017 NEC Fig 690.1(b)]
Courtesy NFPA

Images are good to learn from. Next, we will go over the different images in Figures 690.1(B), paying close attention to the various PV system disconnecting means, which separate the PV system covered here in Article 690 from systems covered in other areas of the 2017 Code. Remember, much of this has changed in the 2017 NEC.

Interactive (grid-tied) inverter circuits are very simple. The inverter is used only for PV power; it has no other purpose and therefore is part of the PV system.

A big question installers have is: "What is the difference between an ac module and a microinverter bolted to a PV module?" The answer is that if the PV module was listed to UL1703 while the inverter was bolted to it and the inverter was tested and listed to UL 1741 while bolted to the PV module, then it is an ac module and we do not consider dc part of the product when installing this module.

> If the module and microinverter were not listed together, then we are responsible for applying the NEC to the dc circuit going from the module to the inverter. It is also interesting to note that the word *microinverter* does not appear in the NEC. The NEC looks at a microinverter as nothing more than a small (micro) inverter.

There is a lot of information in Figure 1.6. First of all, dc coupled and multimode are different things, which can go together. A dc coupled system is a PV system that is typically charging batteries with a charge controller connected to a PV array. The inverter in a dc coupled system will be coupled with the inverter and the charge controller working with dc voltage. In fact, it is possible to have a dc coupled system that does not have an inverter, but most people would like to utilize ac electricity with their dc coupled systems.

As we can see in the 690.2 Definitions that we are about to dive into, a multimode inverter is an inverter that can work in different modes, such as stand-alone (off-grid) and interactive (grid-tied). This type of inverter was also known as a bimodal inverter for a time and will have different outputs. One output will go to the stand-alone (backed up) loads and the other output will go to the loads that are not backed up and to the grid. When the power goes down, the interactive output of the inverter

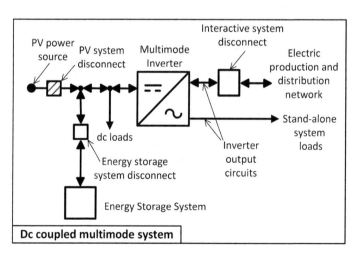

Figure 1.6 Dc coupled multimode system [2017 NEC Fig 690.1(b)]
Courtesy NFPA

will act exactly as an interactive inverter and anti-island (stop sending voltage or current to the grid). No interactive inverter circuit is allowed to be an "island of power" and must disconnect from the grid.

Multimodal vs. hybrid

There is often confusion about multimodal inverters and hybrid PV systems.

A **hybrid** system is defined in Article 100 Definitions as, "A system comprised of different power sources. These power sources could include photovoltaic, wind, micro-hydro generators, engine-driven generators, and others, but do not include electric power production and distribution network systems. Energy storage systems such as batteries, flywheels, or superconducting magnetic storage equipment do not constitute a power source for the purpose of this definition." What we are saying here is that being connected to the grid has nothing to do with being hybrid. Hybrid has to do with having multiple sources of power, not including energy storage or the grid. A **multimodal** system is, as we have mentioned, one that can work in grid (interactive) or off-grid (stand-alone) mode.

Ac coupled systems are becoming more popular. There are arguments on each side, whether it is best to add energy storage to PV systems via ac coupled and dc coupled systems (or both). Ac coupled systems have the benefit of being able to use regular grid-tied inverters in the system and the drawback of having two kinds of inverters.

In Figure 1.7, starting at the upper left, we have a PV array and an interactive inverter, which is the PV system according to the 2017 NEC. On the other hand, according to the 2014 NEC, almost everything in the image is the PV system. We can see that the border that separates the PV system from the rest of the ac coupled multimode system is the PV system disconnect. It is surprising to many that the multimode inverter has no place to connect PV to it. This inverter is connected to an energy storage system (usually batteries) on the dc side and to the grid (electrical production and distribution network) on one ac output and to what I like to call the "ac microgrid" on the other ac circuit, where backed up loads can usually operate. It is also interesting that some manufacturers can make ac coupled systems that will not operate at all when the grid is down. This can be for what is often called "self-consumption" in the industry. These systems will be able to send electricity from the batteries to the loads or the grid when it is

Article 690 photovoltaic (PV) systems 13

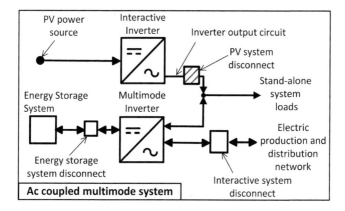

Figure 1.7 Ac coupled multimode system [2017 NEC Fig 690.1(b)]
Courtesy NFPA

beneficial to do so because of utility demand charges, time-of-use rate schedules or because in some places, utility customers are not allowed to export energy.

The stand-alone system in the image above has a few differences with the average stand-alone system we see in the field. First of all, there is usually a charge controller that is connected to three different things. First is the PV array, second is the energy storage system and third is the inverter. These days, it is unusual to have dc loads as shown in Figure 1.8.

Figure 1.8 could also be considered a dc coupled system without a multimode inverter.

690.2 definitions

Because this book is meant to be read with an actual NEC book handy or to be read by someone already familiar with the NEC, we will not repeat every easy to understand definition in Article 690. We will repeat the language of some of the newer and more difficult to understand definitions that a solar professional will have a tendency to use in their career. We will also add discussion to some definitions.

Alternating current (ac) module

Discussion: The question that many solar professionals have is: "What is the difference between an ac module and a microinverter attached

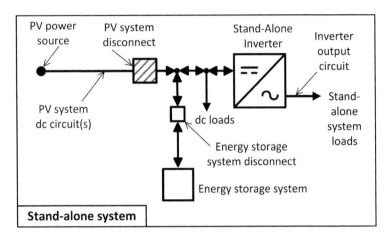

Figure 1.8 Stand-alone system [2017 NEC Fig 690.1(b)]
Courtesy NFPA

to a module?" The answer is that the ac module has the microinverter attached to it before it goes through the UL 1703 PV testing and the UL 1741 inverter testing (also see Figure 1.5, page 10 of this book).

Bipolar photovoltaic array

Discussion: A bipolar PV array is dc power analogy of 120/240Vac power on a house in the US. In a bipolar system there is a positively grounded array section and a negatively grounded array section on the same inverter. This means that we can have voltage to ground that is half of the total voltage that the inverter is getting the benefit of processing. The interesting thing about this, in the 2017 NEC, is that with the 1500Vdc to ground equipment, we can have an inverter with a 3000Vdc input in a ground mount PV system!

This is not something anyone is likely to see in his or her backyard, but according to the 2017 Code, it is a possibility.

At NREL in 2014, Sean and Bill made a proposal to require lithium batteries for bipolar arrays for the 2017 NEC, but everyone just laughed at them.

Article 690 photovoltaic (PV) systems 15

Dc-to-dc converter

Discussion: The dc-to-dc converter definition was put into the Code in 2014 and first put into a diagram in the 2017 NEC. The definition is left rather wide open, so that new equipment not yet in use or invented can be put into use and save lives. For instance, the diagram shows dc-to-dc converters connected to PV source circuits and then the dc-to-dc converters are connected to each other in parallel to make a dc-to-dc converter output circuit. The way we usually see it in practice, at the publication of this book, is with the dc-to-dc converters connected to one PV module per optimizer and then the optimizers are connected together in series and then connected to the inverter. There are some optimizers that work almost the same, but with two modules in series. These optimizers with two modules in series may not comply with module level shutdown requirements that we will learn about when we get to **690.12 Rapid Shutdown**. (In 2019, the 2017 NEC may require module level shutdown unless other products are made available that provide similar safety benefits).

Power optimizers

Dc-to-dc converters are commonly referred to in the industry as "power optimizers," which is really a marketing term. If a dc-to-dc converter did not work as well as advertised or was clipping power (reducing power on purpose), perhaps we would call it a "power-pessimizer."

Dc-to-dc converter source and output circuits

Discussion: Dc-to-dc converter source and output circuits, which are new terms in the 2017 NEC, are like PV source and PV output circuits, however, they are dc-to-dc converters connected together rather than PV modules. Just like how PV modules connected together in series are a PV source circuit, dc-to-dc converters connected together in series are a dc-to-dc converter source circuits.

String theory

We often call a PV source circuit a "string." The term string is not in the NEC but is used in the IEC (International Electrotechnical Commission) and many other international codes and standards. Since everyone in the industry is calling PV source circuits "strings," would it be acceptable to call dc-to-dc converter source circuits "strings"? Perhaps it is as correct as calling a PV source circuit a "string."

On another note, I often hear installers calling microinverters that are connected together on a circuit a "string." I am always correcting the microinverter "string" concept and calling it a "branch" rather than a "string," since microinverters connected together are generally connected in parallel, like a branch circuit, and do not have that "series-string thing" happening.

It is possible for someone to come up with a microinverter technology that requires microinverters to be connected in series with other microinverters that would be lower voltage ac microinverters and the voltages would add.

Dc-to-dc Converter Output Circuits are dc-to-dc Converter Source Circuits connected together in parallel and could be connected together in a **combiner**. Dc-to-dc Converter Output Circuits are uncommon, however, the NEC gurus thought it important enough to define and to let the future decide what circuits will be used.

Diversion charge controller

Discussion: A diversion charge controller will divert charge from a battery to prevent the battery from getting overcharged. A diversion charge controller will often send the excess energy to a diversion load, such as a pump or a heating element in a hot water heater. Also included in this definition of a diversion load is sending energy back to the grid, so if we have a grid-tied battery-backup PV system, when the batteries are charged, we will divert the excess energy to the grid.

Electrical production and distribution network

Discussion: This is a fancy name for a grid, which can be a big utility grid or a smaller micro-sized grid such as a groovy college campus.

Functional grounded PV system

One of the biggest theoretical changes in the 2017 NEC for PV turns everything we formerly thought we knew about PV system grounding upside down or we should say right side up!

From now on, common interactive inverters that are either system grounded via a fuse (old-style formerly known as grounded) and systems that are transformerless (formerly known as ungrounded) and do all their magic through electronics in the inverter are hereby considered **functionally grounded inverters!**

Let's talk about the **"formerly known as grounded"** fuse grounded inverters first (may Prince rest in peace). These inverters are system grounded through a fuse. This means that usually a negative, but sometimes a positive dc busbar in the inverter is connected to the ground busbar via a fuse. When *enough* current goes through the ground wire during a ground fault, then the fuse blows and suddenly the grounded conductor becomes ungrounded. So in this type of system when there is a ground fault, what has been white to indicate it is a grounded conductor may be ungrounded when we are working on it. This previously (no more) required us to have a sign telling us that if a ground fault is indicated, then the grounded conductor may be energized. This is a very confusing label to someone who is not familiar with PV systems. Electricians are used to white-colored grounded conductors to be near ground potential! (Remember: a grounded conductor is a current-carrying conductor that has the same voltage as ground due to a single point of system grounding).

These **"formerly known as grounded"** inverters had a problem. Large PV systems would often have a few milliamps per module of leakage currents taking the parallel fuse pathway home to the negative bus as some of the electrons took the path through glass, frames, cables, insulation and rails when there were no ground faults. As the system became larger, the leakage became greater and the immediate cure was to upsize the ground fault detection and interruption (GFDI) fuse. With a larger GFDI fuse and a foggy day, some systems would not be able to catch a real ground fault. Even without a foggy day a ground fault that was not severe would stay hidden and the system would keep on making good, clean solar energy. Then what would happen in a few cases is the dangerous outcome. A second ground fault would occur and we have what in tennis they call a double fault. An entire array can get short-circuited and start backfeeding through a single string and poof, smoke and the rest we will leave up to your imagination before someone uses this against us in court.

The solution: **"Formerly known as ungrounded"** inverters are also known as transformerless inverters, non-isolated inverters or, as I like to call them, magic boxes. It is easier to call these inverters magic than it is to explain the internal workings of these inverters and to explain the workings of a non-isolated grounded ac circuit. Non-isolated inverters have a reference point to ground related to the internal workings of the inverter, but do not have a grounded conductor. We used to call these inverters European inverters, since having something ungrounded on your house was historically an AHJ no-no in the United States. Then after some modernization, a few puffs of smoke (not a pun) and some explaining, these formerly known as ungrounded inverters soon became the safer norm. After all, they are safer, cheaper and more efficient; safer being the key. These inverters can detect ground faults that are way less than the other inverters. Every day these modern inverters do insulation testing on the PV circuits. They are very sensitive and can detect minor ground faults.

With all of this being said, it is still possible to have a solidly grounded PV array; it's just not common in the modern era. Perhaps you have a small PV system running a water pump. Your PV array negative might be solidly connected to a ground rod. That would be a solid ground and it would be safe to say that the system is more than "functionally" grounded.

> The term functional grounding was taken from Europeans by someone on NEC Code Making Panel 4. "Hey Bill, the Europeans called and they want their functional grounding back!"

Because most of our PV systems are functionally grounded systems, we have a new set of rules that covers all of these functionally grounded systems. No more white negatives (or positives), no more required PV wires (USE-2 is good), no more fuses on two polarities and opening positive and negative in the dc disconnect are required to be functional. We will cover this again as soon as you almost forget it and force it into your long-term memory when we get to the meat of Article 690.

Generating capacity

This is the output of the inverter. This is measured in kW and at 40°C. We often call this our ac system size. **690.7 maximum voltage** and **690.8 circuit sizing and current** have some special exceptions for systems with a generating capacity over 100kW and **Article 691 large-scale photovoltaic (PV) power production facilities** has exceptions that apply to systems with a generating capacity of 5000kW (5MW) or greater. This definition is new in the 2017 NEC.

Interactive system

Discussion: On the street, they call this a grid-tied system.

Perhaps because of the fact that an ac coupled PV system can use interactive inverters, we have to use this fancy term to describe a grid-tied inverter without a grid. In a way, an ac coupled system battery inverter will trick an interactive inverter into thinking there is a grid to turn on the interactive inverter. That is a tricky battery inverter that is not interactive, yet interacting with an interactive inverter to turn it on. We might call this true sine wave inverter coupling love!

Interactive inverter output circuit

Discussion: This term is new in the 2017 NEC and differentiates this circuit from a battery inverter output circuit. Just like it sounds, an interactive inverter output circuit consists of the ac conductors coming out of an interactive inverter. Battery inverter output circuits have different characteristics and now we have different terms for these different circuits.

Multimode inverter

Discussion: A multimode inverter can work in interactive mode or it can work in stand-alone mode. A multimode inverter will have different outputs for the interactive and stand-alone circuits. A multimode inverter has also been called a bimodal inverter in some books. Often multimode inverters are incorrectly called hybrid inverters. Inverters may not be hybrid. A hybrid PV system will have another source of power besides PV, such as a generator.

Solar panels vs. solar modules

Everyone outside of the NEC *realm*, including the President of the United States, calls a solar module the slang term "solar panel." A solar panel according to the NEC is a group of solar modules that are connected together before being installed. In fact, most solar installers today have never installed an NEC defined "solar panel."

Back in the early days of solar in the 20th century, most solar modules had 36 solar cells and were a lot smaller and less powerful than today's solar modules. The solar cells were the expensive part, so it was better for the manufacturers to make something that people can afford. Also, it was common to be charging 12V batteries and 36 solar cells in series is a good design for charging a 12V battery, which is why you see 36-cell modules still around and you see them called 12V solar modules. When 12V modules were what we had to choose from, they were made with screw terminals on the back to attach wires to. These pioneer solar installers would usually put a few modules on a workbench, put some sort of primitive rail behind the modules, connect them together and if they were really good, they would put conduit between the junction boxes of the modules. Conduit between modules was also nice, having no exposed wires on the backs of the modules, so we did not need to put a fence or other structure around the PV system to keep the small fingers from getting in dangerous places. Some commercial PV systems still "panelize" solar modules and connect them together before installing them. I saw this happening in an air-conditioned Phoenix warehouse in the summer, which was a cooler way to do a desert install.

PV source circuit

A PV source circuit consists of PV modules connected together in series, which is commonly referred to as a string.

PV output circuit

If PV source circuits are connected together in parallel at a dc combiner, then the output of the combiner is defined as a PV output circuit.

Discussion: Most PV systems on buildings these days have PV source circuits that go straight to the inverter and avoid having PV output circuits and dc combiners. There are also large GW scale PV systems that use "string inverters" and avoid having PV output circuits and combiners.

Benefits of not having PV output circuits/dc combiners:

1. There is no need for a dc combiner.
2. There are no dc combiner fuses.
3. If one or two strings are going to a single MPP, which is commonly the case, then no dc fuses are needed.
4. Having strings go straight to the inverter puts the dc arc-fault protection at the inverter, which is more convenient than having electronics detect dc arc-faults at dc combiners.
5. Having strings go to the inverter makes monitoring strings more convenient, since we do not need to have monitoring at combiners.

String inverters are becoming more popular at the expense of central inverters for the reasons mentioned above. Some people think that large central inverters will go out of style completely. The authors of this book will remain neutral and let the economics decide.

> ### MPP or MPPT = maximum power point (tracking)
>
> Most modern string inverters have multiple inputs. These inputs are connected to separate dc-to-dc converters in the inverter. This means that different PV source circuits or groups of PV source circuits can operate at the perfect voltage for power production. This way different PV source circuits can have different numbers of modules or bypass contributions from shaded modules through the bypass diodes in the modules. Additionally monitoring and dc arc-fault protection is easier. In the early days before MPPTs, it was a solar sin to have different numbers of modules on different source circuits on the same inverter. Now we have different inverter inputs that can operate independently of each other.

Photovoltaic power source

NEC wording: "An array or aggregate of arrays that generates dc power at system voltage and current."

Discussion: With the advent of dc-to-dc converters and multi-input inverters the definition of Photovoltaic Power Source is more complicated. I would say that with optimizers and microinverters, each set of module conductors can be working at a different voltage and some might view these as separate PV power sources. However, a better way to view these separate circuits is to call them PV source circuits. An inverter with three different source circuits connected to three separate inputs would have no PV output circuit.

Photovoltaic system dc circuit (new in 2017)

NEC wording: "Any dc conductor supplied by a PV power source, including PV source circuits, PV output circuits, dc-to-dc converter source circuits, or dc-to-dc converter output circuits."

Discussion: PV system dc circuits are PV and dc-to-dc converter source and output circuits. A circuit going from a charge controller to a battery or a charge controller to an inverter is not a PV system dc circuit. In fact, as of the 2017 NEC, batteries and charge controllers are not part of a PV system; they are different systems according to the 2017 NEC.

22   Article 690 photovoltaic (PV) systems

Stand-alone system

Synonym: Off-grid or possibly optional standby system (Article 702)
   Note: Stand-alone systems moved from 690.10 to new Article 710 Stand-Alone Systems in the 2017 NEC.

690.4 general requirements

Outline of 690.4 general requirements

690.4 General Requirements

   690.4(A) Photovoltaic Systems
   690.4(B) Equipment
   690.4(C) Qualified Personnel
   690.4(D) Multiple PV Systems
   690.4(E) Locations Not Permitted

One of the more difficult things for someone learning to use the NEC is to remember to know where to look for something. This book is going to do its best to outline, organize, point and discuss topics, so that the reader will be more familiar with and have a better idea where to find what you are looking for.

   Section 690.4 General Requirements, which is in Part I General of the NEC is not very memorable and it is going to stump a few people who are looking for this information, so let us state the obvious and dive into these General Requirements. For whatever reason you can make something catch your attention, it will help you remember it.

690.4(A) photovoltaic systems

In plain English: PV systems can supply a building at the same time as other sources of power.

   Discussion: If you live in coal country and the AHJ refuses to let you put those sunbeam electrons into the grid, you can support your argument here.

690.4(B) equipment

Equipment that needs to be listed (or field labeled) for PV applications according to 690.4(B):

   Inverters (UL 1741)
   Motor Generators [Dc motors driving a rotating generator]

PV Modules (UL 1703)
PV Panels [Products have been built that panelize modules and have been listed when shipped that way. Solyndra solar panels were a group of tubes, each being a module, that were listed and shipped with multiple tubes on a rack creating a solar panel and listed to UL 1703].
Ac Modules [UL 1703 and UL 1741 tested as a unit]
Dc Combiners (UL 1741)
Dc-to-dc Converters (UL 1741)
Charge Controllers (UL 1741)

> Discussion of listed and field labeled equipment: Listed products are found on a list of certified products that various certification labs develop. Field labeled products may not be on one of these lists but get evaluated by a certification lab who puts a label on the product after it has met whatever test requirement was requested to be tested.

690.4(C) qualified personnel

What it means: Installation of equipment and wiring should be done by "qualified personnel."

There is an **informational note** that tells us that we can look to Article 100 Definitions to see the definition of **qualified person**. Article 100 **Qualified Person Definition:** One who has skills and knowledge related to the construction and operation of the electrical equipment and installations and has received safety training to recognize and avoid the hazards involved.

Discussion: Some would say that a qualified solar installer is NABCEP Certified. Others would say only an electrician should install solar and yet others say only a roofer should put a hole in a roof.

> An informational note on informational notes:
>
> Informational notes in the NEC are good ideas, but not requirements. Just like a yellow speed sign tells you it is a good idea to slow down for a corner, an informational note gives you good advice. Informational notes used to be called fine print notes and abbreviated FPN.

24 Article 690 photovoltaic (PV) systems

690.4(D) multiple PV systems

What it means: multiple PV systems are allowed on a single building.

If multiple PV systems on a building are located away from each other, then there must be a directory at each PV system disconnecting means showing where the other disconnecting means are located in accordance with **705.10 directory.**

Discussion: We do not want firefighters thinking they turned off all of the PV on the building when they hit one of the disconnects on the building, not knowing that there are other disconnects that will turn off other PV systems at a different location on the building. Tricking of firefighters or utility workers is not cool nor is it allowed.

Disconnecting means means . . .

Like you would think, a PV system disconnecting means is an off switch for a PV system. A disconnecting means is what separates a PV system from the rest of the electrical system. A PV system disconnecting means for an interactive (grid-tied) inverter would be on the ac side of the inverter, separating the PV system from what is not the PV system. Study PV system disconnect in Figures 1.4 through 1.8 earlier in this chapter on pages 10 through 14 and also in the NEC Figure 690.1(B) Images.

We generally have one PV system disconnecting means and several equipment disconnects for a PV system. Complicated dc PV systems could have more than one PV system disconnecting means, but they have to be grouped for each system.

690.4(E) locations not permitted

PV equipment and disconnecting means are **not allowed in bathrooms** just in case you had your heart set on mounting one next to your toilet – sorry, not allowed.

Think of "wet feet" and getting shocked.

690.6 alternating-current (ac) modules

Outline of 690.6 alternating-current (ac) modules

690.6 Alternating-Current (ac) Modules

    (A) PV Source Circuits
    (B) Inverter Output Circuits

Article 690 photovoltaic (PV) systems

Discussion: 690.6 is stating the obvious.

690.6(A) PV source circuits

What it means: ac modules are tested and listed as a unit, so we do not need to consider any dc circuits, such as PV source circuits.

It is interesting to note that, with a microinverter, we consider the dc conductors between the module and the inverter a PV source circuit, but not with an ac module.

690.6(B) inverter output circuits

It says: The output of an ac module is considered an inverter output circuit.

Discussion: This is obvious, but needs to be explained in case an AHJ gives you a problem.

2 Article 690 photovoltaic systems part II circuit requirements

Part II circuit requirements

690.7 Maximum Voltage
690.8 Circuit Sizing and Current
690.9 Overcurrent Protection [Article 240 is also Overcurrent Protection]
690.10 Stand Alone Systems [moved to Article 710 in the 2017 NEC]
690.11 Arc-Fault Circuit Protection (Direct Current)
690.12 Rapid Shutdown of PV Systems on Buildings [big changes]

690.7 maximum voltage

Understanding 690.7 sets true solar professionals apart from the solar un-professionals. Understanding calculations using 690.7 is also very important to NABCEP, as reflected in their exams.

Outline of 690.7

690.7 Maximum Voltage

    690.7(A) Photovoltaic Source and Output Circuits

        690.7(A)(1) Instructions in Listing or Labeling of the Module
        690.7(A)(2) Crystalline and Multicrystalline Modules
        690.7(A)(3) PV Systems of 100kW or Larger

    690.7(B) Dc-to-dc Converter Source and Output Circuits

        690.7(B)(1) Single dc-to-dc Converter
        690.7(B)(2) Two or More Series Connected dc-to-dc Converters

    690.7(C) Bipolar Source and Output Circuits

Electricians are used to having the grid as the voltage or a device that has a factory set voltage output. With PV, we have a lot of variables.

690.7(A) photovoltaic source and output circuits

PV source and output circuits get their voltage directly from series connected solar cells. The NEC will consider two factors that increase PV source and output circuit voltage. First of all, putting modules in **series increases the voltage**. Secondly, **cold temperature increases the voltage**.

PV output circuits are PV source circuits connected together in parallel at a **dc combiner**. Since voltage is determined by series connections and not parallel connections, **PV output circuits have the same voltage as the PV source circuits that are combined to make the PV output circuit**. From here on out, we will just talk about PV source circuit voltage and understand that the PV output circuit has the same voltage as the PV source circuits that feed it.

Multiple MPPT inverters

Most modern string inverters have multiple Maximum Power Point Trackers (MPPTs). Each MPPT is an electrically separate input that can and will operate at a different voltage from other inputs on the same inverter. Different inputs are electrically treated from a PV designer point of view at the inputs as if they were different inverters. An inverter with two PV series strings per input will not require fuses on the inputs. An inverter with three PV source circuits on one input will typically require fuses.

690.7(A) informational note

An **informational note is a good idea** (not a requirement) and the NEC tells us that a good place to find cold temperature data that we can use in determining voltage for locations in the United States is the ASHRAE Handbook. A very convenient place to find this data is at the website for the Expedited Permit Process: www.solarabcs.org/permitting.

The Solar America Board of Codes and Standards website for the Expedited Permit Process is a document that was put together by Bill Brooks under contract of the United States Department of Energy. On the left side of the www.solarabcs.org webpage, click on Expedited Permit Process and then click on "map of solar reference points" to find the low temperature data to use for calculating voltage. This webpage also has high temperature data that can be used for wire sizing, which we will cover later in this book.

The Expedited Permit Process is a template, which includes fill-in forms that can be used to put together a permit package. Regardless of whether or not you use the templates, there is a lot of good information to study by downloading the 82-page Expedited Permit Process "full report." Anyone in the solar industry will benefit from becoming familiar with this report. It also helps when studying for the NABCEP PV Installation Professional exam.

The 2017 NEC gives us three ways to determine voltage and we can make a choice of which method we will use. These methods will result in different values of voltage, depending on the method we chose to use.

The **three methods** for determining PV source circuit (**string**) **voltage** are:

690.7(A)(1) Calculations
690.7(A)(2) Table 690.7(a)
690.7(A)(3) Engineering supervision

690.7(A)(1) voltage temperature calculation method

The 690.7(A)(1) method is the most common method used by solar professionals for determining PV source circuit (string) voltage. This method is also required for anyone taking any NABCEP PV exam.

In order to calculate the module maximum voltage, you will need three things:

1 Voc (open-circuit voltage)
2 Temperature coefficient of Voc
3 Low temperature

Module **Voc** and **temperature coefficient of Voc** is most commonly found on the PV module manufacturer's datasheet. Low temperature data is most easily found on www.solarabcs.org.

Article 690 photovoltaic systems part II

Let us run through a **PV source circuit maximum voltage calculation** using a simple example with round numbers.

Example:

1. Cold Temp = –5°C
2. Temp. Coef Voc = –0.3%/°C
3. Voc = 40V
4. Number of modules in series = 10

Calculation:

1. Determine **delta T** (difference in temperature) from Standard Testing Conditions (STC)

    a. All PV modules are tested at STC = 25°C
    b. The difference between –5°C and 25°C is 30°C or –30°C

2. Multiply delta T by Temp. Coef. Voc

    a. 30°C × 0.3%/°C = 9% increase in voltage
    b. Another easier method converts percent to decimal first

    　i. 30°C × 0.003 = 0.09

3. Add 1 to figure above to get 109% increase in voltage

    a. 0.09 + 1 = **1.09**

    　i. This figure is a temperature correction factor

4. Multiply the temperature correction factor by Voc at STC to get cold temperature Voc

    a. 1.09 × 40V = **43.6V = maximum voltage** for one module

5. 10 in series × 43.6V = **436V maximum voltage for the PV source circuit** (string)

When practiced, the method above can be done in 10 seconds by fast calculator users. If you practice this 10 times fast, you will be an expert. This method can be done easily with a calculator and without writing anything down.

On the calculator keypad, press:

　25 + 5 = 30 (if the 5°C were above zero then subtract 5 from 30
　　　to get 20)
　30 × .003 = .09

.09 + 1 = 1.09
1.09 × 40 = 43.6V = maximum voltage

Often we do **string sizing** with this number, which means we divide it into the inverter maximum input voltage and then round down to get the **maximum number of modules in series** without going over voltage.

Example using 43.6V maximum voltage and 450V inverter:

450V/43.6V = 10.3

10 in series is the maximum number in series without going over voltage (always round down here).

In this example, if we have 10 in series, then the maximum system voltage is:

10 in series × 43.6V = **436V = maximum system voltage**

It is very common for solar un-professionals to incorrectly write that the maximum system voltage is 450V on the label on the dc disconnect in this example, which is incorrect. 436V is correct.

Once practiced, you should be able to do this calculation without paper using a calculator in less than a minute. The world record is 17.1 seconds.

690.7(A)(2) table method

Table 2.1 NEC Table 690.7(a) voltage correction factors for crystalline and multicrystalline silicon modules

Correction factors for ambient temperatures below 25°C (77°F) (Multiply the rated open-circuit voltage by the appropriate correction factor shown below.)

Ambient temperature (°C)	Factor	Ambient temperature (°F)
24 to 20	1.02	76 to 68
19 to 15	1.04	67 to 59
14 to 10	1.06	58 to 50
9 to 5	1.08	49 to 41
4 to 0	1.10	40 to 32
−1 to −5	1.12	31 to 23
−6 to −10	1.14	22 to 14
−11 to −15	1.16	13 to 5
−16 to −20	1.18	4 to −4
−21 to −25	1.20	−5 to −13
−26 to −30	1.21	−14 to −22
−31 to −35	1.23	−23 to −31
−36 to −40	1.25	−32 to −40

Courtesy NFPA

Using Table 690.7(a) is easier than performing the 690.7(A)(1) Voltage Temperature Calculation Method. We can consider this optional method a shortcut; however, in some cases, we will have more options for more modules in series using the 690.7(A)(1) method. The **690.7(A)(2) method** using Table 690.7(a) is **more conservative** and will come up with a slightly **higher module voltage** every time.

We use **Table 690.7(a)** by cross-referencing a temperature with a **temperature correction factor**. We then multiply the temperature correction factor by the module open-circuit voltage to get the module maximum voltage.

Let us use the **690.7(A)(2) method using Table 690.7(a)**, using the same numbers that we just used in the **690.7(A)(1) calculation** example, however, this time we will not use the module manufacturer's temperature coefficient for open-circuit voltage.

Example:

1. Cold temp = –5°C
2. Voc = 40V
3. Number of modules in series = 10

Calculation using Table 690.7(a)

1. Looking at Table 690.7(a) at –5°C, we can see that –5°C corresponds with a temperature correction factor of 1.12 (12% increase in voltage)
2. Multiply 1.12 × 40V and get 44.8V
3. 10 in series × 44.8V = 448V maximum system voltage

We can see, by the results of comparing the **690.7(A)(1) calculation method** to the **690.7(A)(2) table method**, that the **690.7(a) table method** resulted in a **higher voltage** that is very close to the 450 inverter maximum voltage.

If it were 1 degree colder at –6°C we would still be able to have 10 in series with the **690.7(A)(1) calculation method**, but we would have gone over voltage using the **690.7(A)(2) table method** using Table 690.7(a).

We can see that at –6°C in **Table 690.7(a)** we have a temperature correction factor of 1.14.

40V × 1.14 = 45.6V
45.6V × 10 in series = 456V = over voltage for our 450V inverter example.

32 Article 690 photovoltaic systems part II

690.7(A)(3) engineering supervision method for calculating maximum voltage for PV systems over 100kW generating capacity (ac system size). New in 2017 NEC

Under engineering supervision, there can be alternative ways of doing things throughout the NEC. According to 690.7(A)(3), a **licensed professional electrical engineer** will have to stamp the system design. A professional engineer (PE) has gone to school, worked in the field and taken a difficult exam. The designation of PE is awarded on the state level. It is up to the AHJ to accept a stamp of a PE that is licensed in another state.

The 690.7(A)(3) Engineering Supervision Method requires that the PE uses an "industry standard method" for determining maximum voltage.

690.7(A)(3) informational note

There is an informational note that recommends an "**industry standard method**" for calculating maximum voltage of a PV system.

Industry Standard Method for Calculating Maximum Voltage:

> Photovoltaic Array Performance Model
> Sandia National Laboratory
> SAND 2004–3535
> http://prod.sandia.gov/techlib/access-control.cgi/2004/043535.pdf

To summarize the report: Taking the heating effects of irradiance into consideration, the temperature of the PV will be hotter than ambient and we can have a lower module voltage and perhaps another module in series.

690.7(B) dc-to-dc converter source and output circuits

Dc-to-dc converter source and output circuits shall be calculated in accordance with 690.7(B)(1) and 690.7(B)(2).

690.7(B)(1) single dc-to-dc converter

For a single dc-to-dc converter output, the maximum voltage is the maximum rated output of the dc-to-dc converter.

Maximum voltage is what it says on the label, installation instructions or datasheet for maximum output voltage.

Dc-to-dc converters in modern PV and energy storage systems

As electronics mature, we are seeing dc-to-dc conversion taking place throughout the industry. These are at the inputs of our multiple MPP inverters, MPP charge controllers and within our energy storage systems. One of the reasons that ac became our grid, rather than dc, is because we first had the technology to change voltages with transformers, which work for alternating current and not direct current. Now with efficient modern dc conversion technology, we are seeing more applications for dc circuits.

690.7(B)(2) two or more series connected dc-to-dc converters

Maximum voltage is determined in accordance with instructions of the dc-to-dc converter.

If instructions are not included, sum up the voltage of the dc-to-dc converters connected in series.

Discussion: **Dc-to-dc converters can electronically limit voltage** when connected in series. There are, however, some dc-to-dc converters that, along with having the capability to convert voltages, also have the ability to bypass the converter internally and send PV voltage through the converter. Dc-to-dc converter installation instructions and help lines will be your best source of determining maximum voltage.

690.7(B)(3) bipolar source and output circuits

A bipolar system has a positive grounded array section (subarray) and a negative grounded array section (subarray). Maximum voltage is considered voltage to ground.

If there is a ground-fault or an arc-fault, the inverter is required to isolate both circuits from each other and from ground.

Discussion: From a safety point of view, the voltage of your house is limited to 120V to ground, since voltage to ground relates to safety. In a similar way, a bipolar system with a maximum voltage of 1000V to ground can have a voltage of 2000V measured array to array! However, a 2000V bipolar system is allowed to use 1000V PV modules since the modules never see more than 1000V in this configuration.

34   Article 690 photovoltaic systems part II

Outline of 690.8

690.8 Circuit Sizing and Current

690.8(A) Calculation of Maximum Circuit Current

690.8(A)(1) Photovoltaic Source Circuit Currents

690.8(A)(1)(1) 125 Percent of Short Circuit Current
690.8(A)(1)(2) Systems Over 100kW Engineering Supervision

690.8(A)(2) Photovoltaic Output Circuits
690.8(A)(3) Inverter Output Circuit Current
690.8(A)(4) Stand-Alone Inverter Input Circuit Current
690.8(A)(5) Dc-to-dc Converter Source Circuit Current
690.8(A)(6) Dc-to-dc Converter Output Circuit Current

690.8(B) Conductor Ampacity

690.8(B)(1) Before the Application of Adjustment and Correction Factors
690.8(B)(2) After Application of Adjustment and Correction Factors
690.8(B)(3) Adjustable Electronic Overcurrent Protection Device

690.8(C) Systems with Multiple Direct-Current Voltages
690.8(D) Sizing of Module Interconnection Conductors

690.8 circuit sizing and current

This is how we define current for **wire sizing** and equipment selection. Wire sizing is not simple, so pay close attention and come back often.

690.8(A) and 690.8(B) overview

690.8(A) defines currents used for wire sizing and **690.8(B)** gives us different checks to perform to make sure that the **wire can handle the current under different conditions,** such as heat and continuous current.

There are still **other checks** used for wire sizing regarding overcurrent protection found in Article 240 of the NEC. In **Article 240 Overcurrent Protection** we have to make sure that the **overcurrent** protection device is going to protect the wire. See wiring sizing examples in Chapter 12.

690.8(A) calculation of maximum circuit current

We will **define currents** used for wire sizing for different circuits in a PV system. PV system currents are more complicated than currents in circuits most electricians are used to dealing with. This is because we have **some circuit currents that increase with the brightness** of light and **other** circuit currents that are **limited by smart electronics**.

List of maximum circuit currents:

690.8(A)(1) PV Source Circuits [two methods for defining currents]
690.8(A)(2) PV Output Circuit Currents [based on 690.8(A)(1)]
690.8(A)(3) Inverter Output Circuit Current
690.8(A)(4) Stand-Alone Inverter Input Circuit Current
690.8(A)(5) Dc-to-dc Converter Source Circuit Current
690.8(A)(6) Dc-to-dc Converter Output Circuit Current [based on 690.8(A)(5)]

690.8(A) informational note

The 690.8(A) informational note has confused many prospective solar professionals. It says that when both **690.8(A)(1)** and **690.8(B)(1)** are applied simultaneously, that a **resulting multiplication factor of 156%** is used. Perhaps people find this confusing because the informational note is written before both 690.8(A)(1) and 690.8(B)(1) in the NEC, so **if someone were reading the NEC forward, they would be at a disadvantage**. Often times people try to apply this **156%** factor to circuits besides PV source circuits. This multiplication factor is **only applied** to circuits that are **directly and proportionally influenced by sunlight** and *not* limited by **electronics**. We can call this **"wild PV"** when PV is not limited by electronics.

156% comes from two different 125% correction factors. The **690.8(A)(1)** correction factor is for **natural irradiance beyond the standard testing conditions of 1000W per square meter**, which is how all PV modules are tested and rated. The other 125% in **690.8(B)(1)** is the **required ampacity for continuous current** that we have for all of our solar circuits. **Continuous current** as defined in the NEC is a current that can last over three hours. PV can last all day – particularly in tracking systems!

36   Article 690 photovoltaic systems part II

As a side note, did you know that most silicon solar cells are 156mm × 156mm? Coincidence or photovoltaic numerologist conspiracy?

690.8(A)(1) PV source circuits

There are two new methods for defining PV source circuit (string) currents.

690.8(A)(1)(1) 125% OF SHORT-CIRCUIT CURRENT METHOD

This is the typical way we define maximum circuit current for "wild" PV circuits and is the way we have always done it in the past:

125% of short-circuit current
Isc × 1.25 = maximum circuit current

Discussion: This is the definition of maximum circuit current for one module or a number of modules connected together in series. Some would argue that it is over-doing it to base your wire size on a short-circuit. This is done because we are accounting for increased irradiance.

Also, it is unusual for electricians to see a short circuit that is about 7% greater than operating current, as it is with PV modules.

Let's look at some of the PV module circuit currents in order of increasing current:

Imp = current at maximum power
Isc = short-circuit current
Isc × 1.25 = Maximum circuit current from applying 690.8(A)(1)
Isc × 1.56 = Required ampacity for continuous current from 690.8(B)(1)

Note: Maximum circuit current (Isc × 1.25) is what should be used for the label on a dc disconnect.

It is considered by many smart people to be over-doing it by using 125% of short-circuit current in order to define maximum circuit current. The reason it was not a big deal in the past is because PV was so expensive that oversizing wires was done in order to keep as much of our expensive energy as possible from being lost in the wires. Since we are now entering the age of cheap PV energy, it makes more sense to

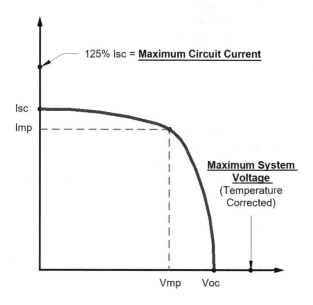

Figure 2.1 IV curve with different currents plotted showing maximum circuit current, which is used for sizing wires, above and beyond short circuit current.

use a smaller wire and this leads us to a new way to define current in 690.8(A)(1)(2), the engineering supervision method.

> **Irradiance outside of Earth's atmosphere**
>
> It is interesting to note that Imp × irradiance in space at Earth's orbit, which is about 1366W per square meter, is about equal to Isc × 1.25.
> Imp × 1.366 = Isc × 1.25
> Coincidence or conspiracy? Clearly the Martians are planning to suck away our atmosphere and want to make sure our electrical code does not need to change after our atmosphere is gone . . .

690.8(A)(1)(2) ENGINEERING SUPERVISION METHOD FOR CALCULATING MAXIMUM CIRCUIT CURRENT FOR PV SYSTEMS OVER 100KW GENERATING CAPACITY (AC SYSTEM SIZE). NEW IN 2017 NEC

As with 690.7(A)(3) for determining voltage, 690.8(A)(1)(2) allows a licensed professional electrical engineer (**PE**) to use an **industry standard** method. This method is based on the **highest three-hour current average** from **simulated local irradiance accounting for elevation and orientation**.

This industry standard method **must also not be less than 70% of the 125% of Isc value used in 690.8(A)(1)(1)**.

Since 690.8(A)(1)(1) is Isc × 1.25 then the industry standard method cannot be less than 70% of 125% of Isc, so 0.7 × 1.25 = 0.875

This means that this **industry standard method cannot be less than 87.5% of Isc**.

We could call this the **not less than 87.5% Isc method**.

This does not mean that the PE does not have to do anything besides multiply Isc × 0.875. The PE will also have to analyze the PV system, including taking irradiance, elevation and orientation into consideration.

Higher elevation causes more current, due to less atmosphere filtering (closer to space).

The Informational Note for 690.8(A)(1)(2) points the professional electrical engineer in the same direction as 690.7(A)(3) and towards the same report mentioned earlier in 690.7 for voltage engineering supervision:

Industry Standard Method for Calculating Maximum Circuit Current:

> Photovoltaic Array Performance Model
> Sandia National Laboratory
> SAND 2004-3535
> http://prod.sandia.gov/techlib/access-control.cgi/2004/043535.pdf

690.8(A)(2) photovoltaic output circuits

PV output circuit currents are the **sum of the parallel-connected PV source circuit currents**.

If you have 10 PV source circuits at the input of a dc combiner, then you will have 10 times the current coming out of the dc combiner through the PV output circuit.

Many PV systems do not have PV output circuits and have PV source circuits that go directly to an inverter.

690.8(A)(3) inverter output circuit current

The maximum current shall be the inverter **continuous** current output **rating**.

Discussion: The current is often marked on the inverter. If current is not marked on the inverter, you can **calculate current by** dividing power by operating voltage. It is interesting to note that dividing power by voltage does not always get the exact same value for current as the inverter datasheet. This can be due to changes in power factor (current and voltage being out of phase with each other and causing a higher current than power divided by voltage would indicate).

Stand-alone inverters often have values for continuous current, which match the power rating of the inverter, but also have **current values for surges of current that are greater than the continuous current rating of the inverter**. For instance, it is not unusual for a 2kW stand-alone inverter to be able to handle a surge of 4kW for several seconds. This is because loads often require surge currents to get started. We size the wires based on the continuous currents and the stand-alone inverter output circuit maximum current, the wires and equipment of which are sized by the continuous current, which is less than the surge currents. **We do *not* size the wires based on the stand-alone inverter's greater surge currents.**

Models:	Sealed FXR2012A
Instantaneous Power (100ms)	4800VA
Surge Power (5 sec)	4500VA
Peak Power (30 min)	2500VA
Continuous Power Rating (@ 25°C)	2000VA
Norminal DC Input Voltage	12VDC
AC Output Voltage (selectable)	120VAC (100-130VAC)
AC Output Frequency (selectable)	60Hz (50Hz)
Continuous AC Output Current (@ 25°C)	16.7AAC
Idle Power	Full: ~34W Search: ~9W
Typical Efficieny	90%

Figure 2.2 Partial datasheet from outback stand-alone inverter
Courtesy Outback Power

Note that in Figure 2.3 there are power ratings that require much more current for time frames that are less than three hours (continuous). Wire sizes are based on currents that are continuous. 16.7A is based on 2000W at 120V. For short periods of time, there will be surges of double the current. Since the time is short, the wire will not have time to heat up and cause a problem.

690.8(A)(4) stand-alone inverter input circuit current

Maximum current shall be the continuous inverter input current when inverter is **producing rated power at lowest input voltage.**

Discussion: A stand-alone inverter takes power from a battery and the **battery will go through a range of voltages,** depending on whether the battery is **fully charged, charging** fast or slow, if the **load is surging** and the **minimum allowable voltage,** before the inverter will no longer take power from the battery in order to protect the battery from too deep of a discharge.

Since power = voltage × current, for consistent power, **as the battery voltage becomes lower, then the current will have to be higher in order to have consistent rated output power.**

For example, with a 1000W inverter, if the voltage is 14V as the battery is charging and we ignore inverter losses, then the required current would be calculated as:

1000W/14V = 71A

As the battery voltage goes down to 11.5A, then the calculation would be:

1000W/11.5V = 87A

If we had a 90% efficient inverter, then we could also calculate for the extra input current required to compensate for inverter losses to be a factor of 0.9.

87A / 0.9 = 97A

Stand-alone inverters **require more input current at lower voltages to make the same amount of power.**

690.8(A)(5) dc-to-dc converter source circuit current

The maximum current shall be the continuous output current rating.

Discussion: Dc-to-dc converters have the ability to electronically control current and voltage. In the instructions of the dc-to-dc converter there should be details on what the maximum current of the circuit can be. One of the benefits of the dc-to-dc converter is its smart ability to limit current, so that smaller wires can be used.

690.8(A)(6) dc-to-dc converter output circuit current

As dc-to-dc converter source circuits are connected together in parallel to make a dc-to-dc converter output circuit, the sum of all of the currents being combined shall be the dc-to-dc converter output circuit current.

Discussion: At the time of the writing of the 2017 NEC, the parallel combining of dc-to-dc converter source circuits is not common, however, it is an option for the future.

690.8(B) conductor ampacity

Conductor Ampacity Code References (abbreviated and interpreted)

690.8(B)(1) Before Application of Adjustment and Correction Factors
690.8(B)(2) After the Application of Adjustment Factors
690.8(A) PV System Current Definitions
690.8(A)(1) PV Source Circuit Current Definition = Usually Isc × 1.25
Table 310.15(B)(2)(a) Ambient Temp. Correction Factors Based on 30°C
Table 310.15(B)(3)(a) Adjustment Factors for More Than Three Current-Carrying Conductors
Table 310.15(B)(16) Ampacities of Insulated Conductors *not* in Free Air
Table 310.15(B)(17) Ampacities of Insulated Conductors in Free Air

---

Note: See Chapter 12 "PV Wire Sizing Examples" for example calculations.

About continuous currents:

> PV system currents shall be considered to be continuous (PV can last more than three hours, since the sun is often up more than three hours per day).

42   *Article 690 photovoltaic systems part II*

Attention! The following is an important part of 690.8(B) that needs to be understood and is commonly misunderstood:
Word-for-word NEC:

> "*Circuit conductors shall be sized to carry not less than the larger of 690.8(B)(1) or (B)(2)* or where protected by a listed adjustable electronic overcurrent protective device in accordance with 690.9(B)(3), not less than the current in 690.9(B)(3)."

The text in bold above needs to be properly understood before going further. We will do a check for 690.8(B)(1) and we will also do a check for 690.8(B)(2), then we will choose the larger wire of the two checks. The wire will always be able to carry more current than the device it is connected to in order to be on the safe side (see Chapter 12 for examples).

690.8(B)(1) before application of adjustment and correction factors

---
Ampacity = Current carrying ability

Here we account for **required ampacity for continuous current** by multiplying the currents we defined in 690.8(A) by 1.25.
Or said more simply:

690.8(A) current × 1.25 = 690.8(B)(1) required ampacity.

Some people call this 690.8(B)(1) value "current." It is not really current; it is a required ampacity that is more than the actual current in order to be extra safe due to current lasting three hours or more (continuous current).

690.8(B)(1) = Required Ampacity for Continuous Current
Terminal announcement regarding 690.8(B)(1) value!

Terminal = what the end (terminal) of a wire is connected to.
Example: screw terminal
Why do we take terminal temperature limits into consideration?
   If a wire is connected to a terminal, there can be some resistance at the connection and the terminal can heat up. The wire can act

> as a heat sink absorbing heat from the terminal. Additionally, if the wire heats up and is connected to the terminal, so will the terminal get hot.

If you open up your NEC to Table 310.15(B)(16) or 310.15(B)(17), you will see values for ampacity for a particular wire that change as you go across the table, depending on the **temperature rating of the insulation** of the conductor. We see 60°C ampacity on the left next to 75°C ampacity and then 90°C ampacity. A conductor that has **an insulation rating that can be hotter will be able to carry more current.**

Terminals have temperature limits and **terminals that are 75°C rated and used with 90°C rated wires are most common in the solar industry.** In general, when doing the **690.8(B)(1)** check, we **use the 75°C column** in the **310.15(B)(16) and (17)** tables in this case.

If we were using 90°C rated terminals with 75°C rated wire, we would use the 75°C column, since it is less than 90°C. It is so uncommon to use 90°C terminals with 75°C wire that we have never seen it done. You can become the first. **90°C terminals are rare on both ends of a circuit.** Both ends of the circuit must have 90°C rated equipment to use the 90°C ampacity at the terminals.

On another note, the NEC tells us that if the terminal temperatures are not indicated, we should assume 60°C terminals if the terminals are rated for 100A or less. This situation of assuming 60°C terminals is probably only seen on an exam. The vast majority of PV equipment has 75°C terminals.

This terminal temperature logic is only used for the 690.8(B)(1) check and not for the 690.8(B)(2) check.

> Terminal Temperature NEC Reference:
>
> Article 110 Requirements for Electrical Installations
> 110.14 Electrical Connections
> 110.14(C) Temperature Limitations
> 110.14(C) reads: "The temperature rating associated with the ampacity of a conductor shall be selected and coordinated so as **not to exceed the lowest temperature rating of any connected termination,** conductor, or device. Conductors with **temperature ratings higher than specified for terminations** shall be permitted to be used for **ampacity adjustment, correction, or both.**"

Essentially, what this is saying is that the terminal is rated for a certain temperature (75°C for instance). The conductor can reach 75°C, but is not permitted to get hotter than 75°C. When we apply conditions of use, we are doing calculations that approximate how much current it takes to get the copper to 75°C. We don't do anything to the terminal. It's just reflecting the conductor temperature.

If we analyze the last sentence in 110.14(C):

"Conductors with **temperature ratings higher than specified for terminations** shall be permitted to be used for **ampacity adjustment**, correction, or both."

We see that we do not need to apply the terminal temperature rating when we are **applying adjustment factors**, which we will read about next.

690.8(B)(2) after the application of adjustment factors

The conductor ampacity should be able to handle the currents as defined in 690.8(A)(1) after the application of correction and **adjustment factors**.

Here are the correction and adjustment factors in the 2017 NEC:

1 Table 310.15(B)(2)(a) Ambient Temp. Correction Factors Based on 30°C
2 Table 310.15(B)(3)(a) Adjustment Factors for More Than Three Current-Carrying Conductors

Additional adjustment in 2014 NEC, but not in 2017 NEC:

3 Table 310.15(B)(3)(c) Ambient Temperature Adjustment for Raceways or Cables Exposed to Sunlight on or Above Rooftops

- In 2017 NEC, if 7/8 inch or less, we still add 33°C to ambient temperature. Table 310.15(B)(3)(c) has been removed in the 2017 NEC.

These 690.8(B)(2) **adjustment factors** are also commonly called "**conditions of use**" since the adjustments have to do with wires put in areas where there will be more heat as with **Table 310.15(B)(2)(a) Ambient Temp. Correction Factors Based on 30°C** or where the wires

Article 690 photovoltaic systems part II   45

will have less of an ability to dissipate heat as with **Table 310.15(B) (3)(a) Adjustment Factors for More Than Three Current-Carrying Conductors.**

Remember that when we apply these adjustment factors, we *do not* apply the criteria in 690.8(B)(1) such as the 125% continuous current calculation or the terminal temperature limits.

TABLE 310.15(B)(2)(A) AMBIENT TEMP. CORRECTION FACTORS
BASED ON 30°C

This table is used to correct (some call derate) the ampacity of conductors in Tables 310.15(B)(16) and 310.15(B)(17) for temperatures that are different from 30°C. Tables 310.15(B)(16) and 310.15(B)(17) are based on temperatures of 30°C, which makes the tables all an electrician has to use if working inside of a building that is not a sauna. When we are working outside of a building in the sun, temperatures can get hotter than 30°C and we can compensate for this by multiplying the conductor ampacity by the derating factor in Table 310.15(B)(2)(a).

> A good place to find high ambient temperatures to use for wire sizing is the www.solarabcs.org expedited permit process on the Expedited Permit Process, map of solar reference points webpage where we also found the cold temperatures that we used for calculating voltage in 690.7. It is recommended to use the temperature value given as the 2% temperature for wire sizing. This is not a record high temperature, but a temperature agreed on by many industry experts.

310.15(B)(3)(A) ADJUSTMENT FACTORS FOR MORE THAN THREE
CURRENT-CARRYING CONDUCTORS

If there are more than three current-carrying wires together in a raceway or cable, then we shall use the derating factors in Table 310.15(B)(3) (a). The reason we do this derating is that the extra conductors in a tight space will generate more heat that has to be dissipated.

We **do *not* count a neutral** that is only **carrying unbalanced currents** from other conductors in the same circuit as a current-carrying conductor. We also **do *not* count equipment grounding** conductors.

46   Article 690 photovoltaic systems part II

TABLE 310.15(B)(3)(C) AMBIENT TEMPERATURE ADJUSTMENT FOR RACEWAYS OR CABLES EXPOSED TO SUNLIGHT ON OR ABOVE ROOFTOPS (REMOVED IN 2017 NEC)

In the 2014, 2011 and 2008 NEC according to Table 310.15(B)(3)(c), we had to add to the ambient temperature a temperature adder if there were conductors in a raceway exposed to sunlight over a rooftop. The reasoning behind this temperature adder is that a conduit in sunlight can act like a solar thermal heater for wires. Apparently, it was not a serious enough consideration to leave the table in the 2017 NEC.

The 2017 NEC still tells us to add 33°C to raceways or cables that are less than 7/8 inch above the roof in sunlight. We recommend installing all conductors more than 7/8 inch above a roof for other reasons besides heat. **Wiring methods close to the roof encourage debris build-up**, which can cause many other problems, such as **roof rot**.

To sum up these **application of adjustment factors,** we take the 690.8(A)(1) defined currents and apply the adjustment factors to determine if the conductor fails to be able to carry the current. Often times with rooftop PV it is this adjustment method that is the weak link and determines the wire size.

We will cover these tables in more detail in our wire-sizing example chapter at the end of the book in chapter 12 on page 189.

690.8(B)(3) adjustable electronic overcurrent protective device (new in 2017 NEC)

Word-for-word NEC:

> 690.8(B)(3): "The rating or setting of an adjustable electronic overcurrent protective device installed **in accordance with 240.6.**"

What does it mean?

240.6(B) Adjustable-Trip Circuit Breakers

Adjustable trip breakers **without restricted access** shall be considered the **highest trip rating possible.**

Discussion: This is because unqualified personnel may set the trip point of unrestricted access adjustable-trip circuit breakers to the highest setting, offering the least protection.

240.6(C) **Restricted Access** Adjustable-Trip Circuit Breakers

Restricted access adjustable-trip circuit breakers shall be permitted to have the **ampere rating equal to the adjusted trip setting.**

Discussion: Since access is restricted to qualified persons, we do not expect someone to come along when the breaker trips and adjust for higher current, leaving the circuit unprotected.

690.8(C) systems with multiple direct-current voltages

If a PV power source has **multiple output circuits** with **multiple output voltages** and employs a common return conductor, the ampacity of the **common return conductor** shall not be less than the sum of the ampere ratings of the overcurrent devices of the individual output circuits.

If we use a single common return wire for three circuits, the return wire must be capable of handling the current of all three circuits. We used to see this done with 48-volt PV systems connected to batteries.

690.8(D) sizing of module interconnection conductors

Word-for-word NEC:

> "Where a **single overcurrent device** is used to protect a set of **two or more parallel-connected module circuits**, the ampacity of each of the module interconnection conductors shall **not be less than** the **sum** of the rating of the single **overcurrent device plus 125 percent of the short circuit current** from the *other* **parallel-connected modules.**"

The PV module parallel-connected circuit shown in Figure 2.4 is most often used with low current thin film PV modules in Utility Scale projects using central inverters.

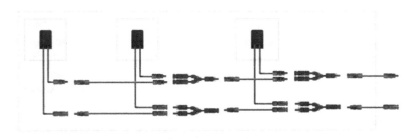

Figure 2.3 Module interconnect for multiple parallel-connected module circuits
Courtesy Shoals

If we have 10 modules with an Isc rating of 1A each, we could imagine a short circuit or a fault on one of the modules, with the currents of all the other modules backfeeding through the fault or short circuit.

We will also figure that the maximum series fuse rating for this module and interconnect conductors is a minimum of 15A. The maximum series fuse rating of a module is like the ampacity of the module.

This would mean nine modules of backfeed current through the shorted or faulted module interconnect and the calculation and checks would be as follows:

Current = 1A × 1.25 × 9 = 11.25A >> minimum fuse is 15A.
Total current available to the module is 11.25A + 15A (fuse) = 26.25A (fail)

- (15A fuse currents are from different part of array)

Reduce guess to six modules in parallel.

Current = 1A × 1.25 × 6 = 7.5A >> 10A fuse.
Total current available to module is 7.5A + 10A= 17.5 >> Pass.
Check 7 in parallel. Current = 1A × 1.25 × 7 = 8.75A >> 11A.
Total current available to module is 8.75A +11A = 19.75A >> Pass with 11A
fuse and no more than seven modules in parallel.

Now wasn't that fun? With the new definitions of current we can use even lower numbers for module current based on simulations for systems over 100kW.

Discussion: Some PV systems have modules with low current and higher voltage. These modules are sometimes connected in parallel to a single overcurrent protection device. To some people, it would look like two (or more) PV source circuits combining at a single fuse. In fact, these two series connected groups of modules are in fact considered a single PV source circuit. This is why in 690.8(A)(1) Photovoltaic **Source Circuit** Currents method (1), it says "the sum of **parallel-connected** PV module rated short-circuit currents multiplied by 125 percent."

Yes, that is right; **a PV source circuit can be made of two** or more **"strings" in effect** going to a single fuse as long as the module interconnection conductors can handle the currents from the number of strings minus one. We subtract one because the string being back fed on by the other strings in a short circuit condition does not need to be counted.

This logic is also used in the case that most designers know about in which two PV source circuits combining at a single MPP do not need a fuse.

690.9 overcurrent protection

Outline of 690.9

690.9 Overcurrent Protection

    690.9(A) Circuits and Equipment
    690.9(A) Exception: overcurrent protection not required
    690.9(A) Exception (1): no external sources
    690.9(A) Exception (2): short-circuits do not exceed ampacity
    690.9(A) Exception, Informational Note: current-limited
    690.9(B) Overcurrent Device Ratings

        690.9(B)(1) Not less than 125% of maximum currents
        690.9(B)(2) 100% of its rating
        690.9(B)(3) Adjustable electronic overcurrent protective devices
        690.9(B)(3) Informational Note: adjustable OCPD preventing backfeeding

    690.9(C) Photovoltaic Source and Output Circuits
    690.9(C) Informational Note: OCPD only in positive or negative
    690.9(D) Power Transformers
    690.9(D) Exception: Side of transformer towards inverter

Article 690.9, PV overcurrent protection follows the line with **Article 240 Overcurrent Protection,** but with special provisions for PV that are different from most electricity, such as solar cells that produce current based on the brightness of light, the current-limited aspects of PV and current that can be flowing in different directions.

690.9(A) circuits and equipment

PV system dc circuit and ac inverter output conductors and equipment shall be protected against overcurrent.

Overcurrent protective devices are not required for circuits with sufficient ampacity for highest available currents.

50 *Article 690 photovoltaic systems part II*

Examples of **current-limited supplies:**

- PV modules
- Dc-to-dc converters
- Interactive inverter output circuits

Circuits connected to current-limited supplies and also connected to sources with higher current availability shall be protected at the higher current source connection.

Examples of higher current availability:

- Parallel **strings** of modules
- Utility power

Strings

690.9(A) is the only place in Article 690 where it says string or strings rather than PV source circuit. String is a term that is defined in the International Electrotechnical Commission (IEC) standard. We often call a PV source circuit a "string" and this can increase the use of slang.

The IEC is an international standard that is used to align Codes around the world. At the beginning of the NEC, on the first page of code, we can see:

> *NEC 90.1(C) Relation to Other Standards*
> *The requirements of this Code address the fundamental principles of protection for safety contained in Section 131 of International Electrotechnical Standard 60364–1 Electrical Installations of Buildings*

This means that the NEC follows the principles of the international standard that most places on Earth attempt to follow to a degree.

The NEC goes into much more detail regarding PV systems than the IEC.

IEC Definition:

> PV string: A circuit of series connected modules

690.9(A) exceptions: overcurrent protection not required

There are instances where overcurrent protection, if used, would create a false sense of security and an unnecessary source of nuisance failures (fuses fail even when they don't get overcurrent at times). In these cases, overcurrent protection is not required.

Since PV itself is current-limited and we size fuses at a minimum of 156% of short circuit current, in many cases a short circuit will not open a fuse.

Let us look at some of these current-limited cases.

690.9(A) exception (1) no external sources

If there are no external sources, such as parallel-connected PV source circuits, batteries or backfeed from inverters.

Let's give an example: If there is a single string of PV connected to an inverter and there is a short circuit of the PV source circuit we would never open up the circuit with an overcurrent protection device. **PV source circuit fuses are sized based on 156% of short circuit current and then rounded up to the next common overcurrent protection device.** If a PV module short-circuit current (Isc) is 9A, then the fuse size we would use would be calculated as:

9A × 1.56 = 14A then round up to 15A fuse

We round up to the next common overcurrent protection device size. Common sizes are found in NEC 240.6. (800A and over we do not round up).

If we had a single string of 9A Isc PV modules short out, we would never get enough current to open a 15A fuse.

In fact, if we had a single PV output circuit short out where it was connected to an inverter, it still would not have enough current to open any properly sized overcurrent protection device on the PV output circuit. We could short-circuit a MW of PV for 20 years and never blow a fuse!

This is the nature of current-limited PV. In some cases it is safer, since we are not exposed to super high short circuits. The arc-flash danger is less, but it can be more dangerous in some aspects, since we cannot open up overcurrent protection devices with large enough currents in many cases.

690.9(A) exception (2) short-circuits do not exceed ampacity

If short-circuit currents from all sources do not exceed the ampacity of the conductors and the maximum overcurrent protective device size rating specified for the PV module or dc-to-dc converter, then we do not need to have overcurrent protection.

This is the example of not needing fuses when you have two PV source circuits going to a single inverter input. If we have one PV source circuit that is shorted out and the currents from the other PV source circuit are backfeeding to the shorted PV source circuit, then we would only have a maximum of the current from a single PV source circuit. We would not have the currents from both strings, since we would at most have currents from one PV source circuit feeding to another. In this case we do not need a fuse.

Why do we have PV source circuit fuses?

If we had 50 PV source circuits combining at a dc combiner and a single string shorted, then we would get the current from 49 PV source circuits backfeeding a single PV source circuit. The fuse protecting the single PV source circuit would have currents going the reverse direction and would open up the fuse, even on a cloudy (low current) day. Dc combiner fuses are designed to open due to currents going in the reverse direction when the non-shorted PV source circuits send current back through a shorted PV source circuit.

690.9(A) exception, informational note: current-limited

PV dc circuits are current-limited circuits that only need overcurrent protection when connected in parallel to higher current sources. The overcurrent device is often installed at the higher current source end of the circuit.

Discussion: We install fuses at the source of the overcurrents at **dc combiners**. Some think it would be easier if there were fuses that came in the junction boxes of PV modules. The source of the overcurrents is not from the individual PV modules, but from the parallel-connected sources feeding back to the shorted PV source circuit. **Overcurrent protection should be installed at the source of the potential overcurrents,** which is where the PV source circuits are combined, at the dc source circuit combiner.

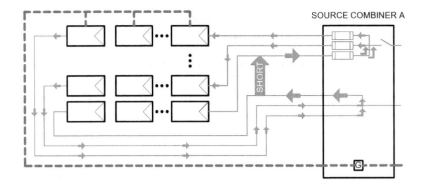

Figure 2.4 Two PV source circuits backfeeding a short on another PV source circuit

Courtesy Robert Price AxisSolarDesign.com

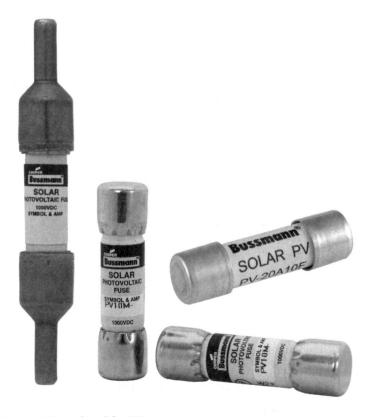

Figure 2.5 Fuses listed for PV

Courtesy Cooper Bussmann

690.9(B) overcurrent device ratings

Overcurrent protection devices used in PV dc circuits shall be **listed** for use in PV systems.

The requirement for a "listed" PV fuse was brought into the NEC in 2014.

690.9(B)(1) not less than 125% of maximum currents

Overcurrent protection devices must be **at least 125% of the currents defined in 690.8(A) Maximum Circuit Current.** You can review those currents defined in 690.8(A)(1) through 690.8(A)(2) starting on page 36 of this book.

690.9(B)(2) 100% of its rating

If an assembly and its overcurrent protective device is rated for continuous use at **100% of its rating,** then it does *not* need to be at least 125% of its rating.

Invoking this "100% of its rating" section of the Code is rare, used for larger currents and not something most designers of smaller PV systems should concern themselves with.

690.9(B)(3) adjustable electronic overcurrent protective devices

We can use adjustable electronic overcurrent devices. There are the same requirements for restricted and non-restricted access that we discussed on page 46 of this book when we were discussing **690.8(B)(3) Adjustable Electronic Overcurrent Protective Device.**

Restricted access means we can rely upon the setting and un-restricted access implies that we have to consider the highest possible setting.

690.9(B)(3) informational note: adjustable OCPD preventing backfeeding

This is a note reminding us that some adjustable overcurrent protective devices can identify and prevent backfeeding.

690.9(C) photovoltaic source and output circuits

A single overcurrent protective device, where required shall be permitted for each PV source circuit or each PV output circuit.

When single overcurrent protection devices are used, they should all be in the same polarity.

Discussion: In previous versions of the NEC transformerless inverters that were formerly called "ungrounded" inverters, which is the typical interactive inverter used today, were required to have overcurrent protection on both positive and negative polarities. The **2017 NEC changes that and only requires overcurrent protection on a single polarity**, when overcurrent protection is required. Now all PV source circuits and PV output circuits require only overcurrent protection on one polarity.

690.9(C) informational note: OCPD only in positive or negative

Due to improved ground fault protection requirements, a **single overcurrent protection device** in either positive or negative in combination with ground-fault protection **provides adequate overcurrent protection**.

690.9(D) power transformers

Overcurrent protection for a transformer with sources on both sides shall consider one side of the transformer, then the other side as primary.

Discussion: Transformer overcurrent protection shall be done in accordance with 450.3. Section 450.3 has different values that overcurrent protection is based on, which are often more than 125% of current.

Article 450 is Transformers and Transformer Vaults and Section 450.3 is for Overcurrent Protection for transformers (not conductors).

Article 240 is what we refer to for overcurrent protection of conductors.

690.9(D) exception: side of transformer towards inverter

If the current rating of the interactive inverter side of the transformer is at least the current rating of the inverter, then **overcurrent on the inverter side of the transformer is not required**.

Discussion: If the transformer can handle all of the current from the inverter, then overcurrent protection is not required.

This is due to the **current-limited characteristics of the inverter** not being able to hurt the transformer or overcurrent the conductors connected between the inverter and the transformer.

690.10 stand-alone systems

Wiring of stand-alone systems shall be done in accordance with 710.15.

- Article 710 is Stand-Alone Systems.
- Section 710.15 General is 99% of Article 710.

Discussion: 690.10 Stand-Alone Systems was big in the earlier versions of the NEC and the material has just moved to a new article, Article 710 Stand-Alone Systems in the 2017 NEC. This is because the definition of a PV system has changed in the 2017 NEC. No longer is energy storage or loads considered part of a PV system and this information has all moved to different locations in the NEC. Much of the requirements are the same, they are just found in a different location. We cover this material on page 173.

690.11 arc-fault circuit protection (direct current)

PV systems 80V or greater shall be protected by a listed arc-fault circuit interrupter.

690.11 arc-fault circuit protection exception

PV systems that are *not* on or in buildings, *PV output circuits and dc-to-dc converter output circuits* that are direct buried, in metallic raceways or in enclosed metallic cable trays are **permitted** *without* **arc-fault circuit protection.**

Buildings that have the sole purpose of housing PV equipment are not required to be considered a building for this exception.

This exception only applies to PV and dc-to-dc converter output circuits. PV source circuits and dc-to-dc converter source circuits are still required to have dc arc-fault protection.

See page 136 for exceptions for dc arc-fault protection under engineering supervision for large (greater than 5MWac) systems in Article 691 Large-Scale Photovoltaic (PV) Electric Power Production Facility.

Discussion and history:

> 2011 NEC 690.11 **only applies to circuits penetrating or on a building** and was the first time 690.11 appeared in the NEC.
> 2014 NEC 690.11 **applies to all dc PV circuits**, even large ground mounts. This is a problem for large central inverter PV systems

Figure 2.6 Dangerous dc arc-fault (do not try this at home)

Smoking kills

Photo by Sean White's solar powered lighter company

that would require dc arc-fault protection for PV source circuits and **PV output circuits.**

Engineering dc arc-fault protection is close to impossible for large PV output circuits. Some would say that utility scale PV using large central inverters does not comply with the 2014 NEC, since there is no way to perform dc arc-fault protection on the PV output circuits.

2017 NEC allows PV systems that are *not* **on or in buildings to not have dc arc-fault protection on PV output circuits when those circuits are either buried or in metallic raceways.** PV source circuits still require arc-fault detection if they are greater than 80V.

Part II of Article 690 is almost complete. The last section of Part II will be covered in the next chapter.

3 Section 690.12 rapid shutdown

Section 690.12 Rapid Shutdown is the hottest spot in PV education since it came out in the 2014 NEC. Chapter 4 of this book is dedicated to covering the evolution of 690.12. Not only are there changes from the 2014 NEC to the 2017 NEC, but also there are parts of the 2017 NEC that will not take effect until 2019. Therefore 690.12 deserves its own chapter.

Overview

The rapid shutdown requirements of the NEC first appeared in the 2014 NEC and have changed quite a bit in the 2017 NEC. This evolution of the Code has a **purpose: to save firefighters' lives**. Additionally, in making houses with PV safer for firefighters, firefighters will be more inclined to save buildings that have PV on them. Having firefighters save buildings with PV on them has many benefits, including keeping insurance rates from going up for buildings sporting PV arrays.

Outline of 2017 NEC 690.12

690.12 Rapid Shutdown of PV Systems on Buildings
690.12 Exception

    690.12(A) Controlled Conductors
    690.12(B) Controlled Limits

        690.12(B)(1) Outside the Array Boundary
        690.12(B)(2) Inside the Array Boundary [effective 2019]

            690.12(B)(2)(1) Listed or field labeled [effective 2019]
            690.12(B)(2)(2) 30 seconds, 80V [effective 2019]

Section 690.12 rapid shutdown 59

690.12(B)(2)(3) No exposed wiring or metal [effective 2019]

690.12(C) Initiation Device

690.12(C)(1) Service disconnecting means
690.12(C)(2) PV system disconnecting means
690.12(C)(3) Readily accessible switch

690.12(D) Equipment

Outline of 2014 NEC 690.12

690.12 Rapid Shutdown of PV Systems on Buildings

690.12(1) Controlled conductors 5' inside 10' outside
690.12(2) 30V, 30 seconds [10 sec. changed to 30 sec. effective 2016]
690.12(3) Voltage measured between any two conductors and ground
690.12(4) Initiation method labeled in accordance with 690.56(B)
690.12(5) Equipment listed and identified

In comparing the outlines of the 2017 and 2014 690.12 rapid shutdown requirements, we can see that **690.12 has been totally rewritten**. Additionally, there are mid-cycle changes taking place with regards to 690.12.

There was a **TIA (Tentative Interim Amendment)** to 2014 690.12(2), which **changed** the **time required for rapid shutdown** of the controlled conductors from **10 seconds to 30 seconds** and went into effect in August of 2016. This change was enacted in order to make the shutdown time for the 2014 NEC match the 2017 NEC and to make it more straightforward for the product manufacturers who build products to meet both code requirements.

The 2017 NEC has some requirements that **do not take effect until 2019**. These requirements are the 690.12(B)(2) **inside of the array** rapid shutdown requirements.

We will organize the rest of this chapter by following the outline of the 2017 NEC requirements, but we will still be comparing and contrasting to the 2014 NEC.

690.12 rapid shutdown of PV systems on buildings

PV systems in or on buildings shall have a rapid shutdown function system to reduce shock hazards for emergency responders.

690.12 exception

If a building's sole purpose is to house PV system equipment, then it does not have to comply with 690.12.

690.12(A) controlled conductors

The requirements for the controlled conductors shall be applied to PV **circuits supplied by the PV system.**

Discussion: Controlled conductors are conductors that we can turn off or control with our rapid shutdown disconnect.

The 2017 NEC has new and reduced definitions of what a PV system is, which we discussed in chapter 1 of this book. These changes **remove the energy storage system and loads** from the definition of all PV systems. We can see this in the images in the 2017 NEC 690.1(b) figures and in chapter 1 of this book beginning on page 10. The **PV system disconnect in each figure is the boundary of the "PV system."** In the 2014 NEC, rapid shutdown requirements applied to battery systems and stand-alone inverters. The 2017 NEC's rapid shutdown requirements only apply to our new and reduced definition of a PV system, which does not include energy storage systems.

690.12(B) controlled limits

We have different rules for inside vs. outside of the array boundary. The **array boundary** as defined in the 2017 NEC is **1 foot from the array.**

We can see the **690.2 definition of Array:**

> **Array.** A mechanically integrated assembly of module(s) or panel(s) with a **support structure** and **foundation**, tracker, and **other components**, as required, to form a dc or ac **power-producing unit.**

Discussion: The array boundary is usually going to be the edge of the PV, but if the rails, tracker or concrete foundation stick out more than the PV, then the array boundary can be 1 foot from the edge of it all.

In the 2014 NEC, according to **2014 NEC 690.12(1)**, controlled conductors are required 5 feet **in length** from the array inside of the building or 10 feet **in length** from the array.

A big difference in the 2014 vs. the 2017 NEC is that the **2014 NEC measures the length of the conductor,** so if you ran a conductor 20 feet long that was 6 inches from the edge of the array, it would not comply with the 2014 NEC. This same **20-foot-long conductor that is run 6 inches from the edge of the array would qualify as inside the array boundary according to the 2017 NEC.**

690.12(B)(1) outside the array boundary

Controlled conductors outside the array boundary or **more than 3 feet from the point of entry inside a building** shall be limited to no more than **30V within 30 seconds** of rapid shutdown initiation.

Voltage shall be measured between any two conductors and between any conductor and ground.

Discussion: First of all, we point out that besides having a limit of 1 foot from the array, **the 2017 NEC also gives us 3 feet from the point of entry inside the building** to have controlled conductors, which can be more than 1 foot from the array boundary if inside a building.

This is interesting, because the 2014 NEC had a 10-foot requirement outside the building and a 5-foot requirement inside the building. In the 2017 NEC the **longer distance changed to inside**, however, all distances were reduced.

The reason that we need at least 3 feet inside the building is to allow for enough conductor to mount equipment inside an attic space or similar area. The roof thickness can take up much of that distance. Most of the time the shutdown devices will be on the roof. Building-integrated PV systems are one example where this 3-foot rule will be important. Additionally, firefighters will probably not be cutting directly through the building under the PV while the PV system is energized.

Controlled conductors **outside of the array boundary** or **within 3 feet from the penetration** of a building must be limited to

- 30V
- Within 30 seconds

In the **2014 NEC, 690.12(2)** first stated that controlled conductors shall be limited to not more than **30V and 240 volt-amperes within 10 seconds**. The 2014 NEC 690.12(2) was **changed in 2016** from **10 seconds to 30 seconds via a TIA.** The reason for this change was to allow product manufacturers to address grid support requirements that may require that the PV array stay on for up to 20 seconds during utility grid problems. It also allows more time for the capacitors on the dc side of the inverter to discharge in 30 seconds. The fire service also agreed that the danger to firefighters would not change significantly with the time increase.

62 Section 690.12 rapid shutdown

The 2014 NEC 240 volt-ampere requirement was also taken out of the code, since it is nearly impossible to verify in the field and needs to be verified by the laboratories that certify the equipment that is used for rapid shutdown.

690.12(B)(2) inside the array boundary (effective 2019)

The PV system shall comply with one of the following:

1. 690.12(B)(2)(1) Listed or field labeled [effective 2019]
2. 690.12(B)(2)(2) 30 seconds, 80V [effective 2019]
3. 690.12(B)(2)(3) No exposed wiring methods [effective 2019]

690.12(B)(2)(1) LISTED OR FIELD LABELED

At the time of the writing of this book, there is no such thing as a listed or labeled rapid shutdown array. Initial research is being performed to help inform what this new standard will contain, but the standard will likely not be available until the 2019 effective date. Products meeting this new standard will also take time to come to market.

690.12(B)(2)(2) 30 SECONDS, 80V [EFFECTIVE 2019]

AKA Module Level Shutdown

690.12(B)(2)(2) is what everybody is talking about. It amounts to **module level shutdown,** which is already very **common with microinverters and dc-to-dc converters** (AKA power optimizers). There are products where the whole array operates below 80V, but these systems are far less common than power optimizers and microinverters.

The reason that 80V amounts to module level rapid shutdown is because the maximum voltage of two 60 or 72 cell modules in series would be over 80V. It is uncommon to find a PV module for a building that would have a maximum voltage over 80V. Even rare 96 cell module will remain under 80V on a cold day in New Jersey. Some thin film modules are over 80V and may not be usable to meet this particular compliance option.

690.12(B)(2)(2) mentions that besides conductors being inside the array boundary having to adhere by 690.12(B)(2)(2), also conductors not more than 3 feet from penetration of the building have to comply. This is because the **690.12(B)(1) definition of outside the array boundary includes within 3 feet of penetration of the building.** Perhaps we could say that 3 feet from the penetration of a building qualifies as inside of the

Section 690.12 rapid shutdown 63

array boundary for this purpose. The reason for this 3-foot rule was to make it clear that the wiring inside the building, immediately under the array, could not be counted as being within the 1-foot array boundary.

This 690.12(B)(2)(2) method is accomplished with electronics distributed throughout the array for the most part. However, this method can also be accomplished with parallel connections and special equipment that disconnect series connected modules at the module level.

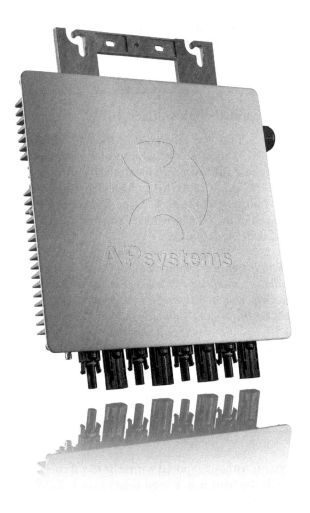

Figure 3.1 AP system 4 module inverter
Courtesy AP Systems

64 Section 690.12 rapid shutdown

690.12(B)(2)(3) NO EXPOSED WIRING OR METAL

PV arrays with no exposed wiring methods, no exposed conductive parts and installed more than 8 feet from exposed grounded conductive parts shall not be required to comply with 690.12(B)(2).

Discussion: To meet the requirements of 690.12(B)(2)(3), we can envision a PV array that has **no exposed metal or conductors** and is **installed at least 8 feet from grounded metal on a roof.**

An example of a PV system that could comply with 690.12(B)(2)(3) would be a Building Integrated PV (BIPV) system with no exposed metal parts, since many BIPV systems do not have metal racks or frames.

Building integrated photovoltaics (BIPV)

The BIPV definition was taken out of the 690.2 Definitions in the 2017 NEC. This was not because of a conspiracy against BIPV, but because BIPV was not mentioned anywhere else in the NEC. The NEC is not supposed to have definitions that do not refer to anything else in the NEC. Other commonly used devices not mentioned in the NEC are microinverters.

2014 NEC BIPV definition:

> **Building Integrated Photovoltaics.** *Photovoltaic cells, devices, modules, or modular materials that are integrated into the outer surface or structure of a building and serve as the outer protective surface of the building.*

It is interesting to note that **Building Integrated Photovoltaics** is still in the index of the 2017 NEC and refers to 690.2 Definitions because someone forgot to take it out of the index. This is also the only place in the NEC where the word photovoltaics is used, rather than photovoltaic.

Any PV system without exposed wiring or metal could comply with 690.12(B)(2)(3). Perhaps it is a good time to come up with your 158 million-dollar idea.

Once again, none of the three 690.12(B)(2) requirements are effective until 2019 according to the 2017 NEC.

690.12(C) initiation device

The rapid shutdown initiation device when in the off position shall indicate that rapid shutdown has been initiated.

Where multiple PV systems with rapid shutdown are installed on a single service, the initiation devices shall consist of no more than six switches or circuit breakers in a single enclosure or group of enclosures.

Where auxiliary initiation devices are installed, they shall control all PV systems with rapid shutdown on that service. An example of an auxiliary initiation device could be a fire alarm system. It is possible to use a relay in a fire alarm system to automatically shut down a PV system even if the main initiation device is not turned off. The key here is that if we are going to call something a rapid shutdown switch, it has to do it all – not just a portion of the system. Any switch that only shuts down part of a system is simply a maintenance disconnect.

Following are the **three methods** used for rapid shutdown initiation:

690.12(C)(1) service disconnecting means

The service disconnecting means (main breaker) can be the rapid shutdown initiation device.

690.12(C)(2) PV system disconnecting means

The PV system disconnecting means (usually a circuit breaker or fused disconnect) can be a rapid shutdown initiation device.

690.12(C)(3) readily accessible switch

A readily accessible switch (often a special rapid shutdown switch or inverter disconnect) can be the rapid shutdown initiation device, shown in Figure 3.2.

690.12(C)(3) informational note

One reason why a 690.12(C)(3) readily accessible switch type initiation device is used is for systems that operate with an optional standby mode and keep operating upon loss of utility voltage.

Discussion: Many rapid shutdown systems shut down whenever the utility shuts down. For systems with backup, there needs to be a way

Section 690.12 rapid shutdown

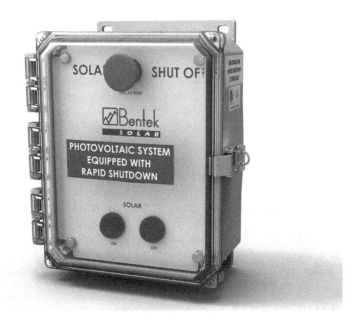

Figure 3.2 Rapid shutdown initiation switch
Courtesy Bentek Solar

to initiate rapid shutdown independent of utility outages. Otherwise, the rapid shutdown function would negate the benefit of the backup power system.

690.12(D) equipment

Equipment that performs rapid shutdown, other than initiation devices, such as disconnect switches, circuit breakers or control switches, shall be listed for providing rapid shutdown protection.

Discussion: The 2014 NEC did not require special listed rapid shutdown equipment. When disconnecting conductors on a roof, using the 2014 NEC, we could use equipment such as listed contactors for PV systems (remote controlled switches) to turn off a system on the roof, which we also used for other purposes besides rapid shutdown of PV systems. Now, according to the 2017 NEC, our rapid shutdown equipment needs to be special listed for rapid shutdown.

We can still use regular switches and breakers to provide rapid shutdown initiation, such as the main breaker or a breaker in a sub-panel.

Section 690.12 rapid shutdown

The code recognizes that it would be ridiculous to get a special listing for a switch when it is being used as just a switch. However, some switches may have special communications functions like the Midnite Solar "birdhouse." Specialized switches like that product may require special listing for rapid shutdown.

Rapid shutdown labeling from 690.56(C)

This chapter is going to include the labeling requirements for rapid shutdown systems which are found in **690 Part VI Marking / 690.56 Identification of Power Sources**. The other question here is: Is this book going out of order, or is the NEC out of order? For discussion purposes, it is a good idea to put all of the rapid shutdown material together.

Outline of 690.56(C)

690.56 Identification of Power Sources

    690.56(C) Buildings with Rapid Shutdown

        690.56(C)(1) Rapid Shutdown Type

            690.56(C)(1)(a) Array level shutdown and conductors leaving array
            690.56(C)(1)(b) Conductors leaving array level shutdown

        690.56(C)(2) Buildings with More Than One Rapid Shutdown Type
        690.56(C)(3) Rapid Shutdown Switch

690.56(C) rapid shutdown type

Buildings with PV systems shall have permanent labels as described in 690.56(C)(1)(a) array level rapid shutdown or 690.56(C)(1)(b) conductors leaving the array level shutdown.

What we are looking at with labeling is differentiating the different types of rapid shutdown. The big difference is whether we have rapid shutdown inside the array boundary, or if there are energized conductors within the array.

690.56(C)(1)(a) shutdown inside the array and conductors leaving array

690.56(C)(1)(a) refers to the highest level of rapid shutdown, where we reduce hazards inside the array.

68 *Section 690.12 rapid shutdown*

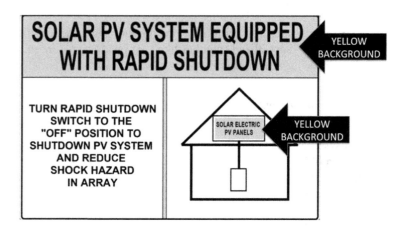

Figure 3.3 NEC Figure 690.56(C)(1)(a) reduced array shock hazard sign
Courtesy NFPA

There are specific requirements for a sign that includes specific colors and words. As a note, the NEC does not have the colors in the example of the sign and neither do the pages of this book.

The scheme for the 690.56(C)(1)(a) rapid shutdown sign is:

Title wording:

SOLAR PV SYSTEM IS EQUIPPED WITH RAPID SHUTDOWN

Title characters:

Capitalized black 3/8 inch minimum height

Title background color:

Yellow (Yellow is the ANSI caution color)

Remaining characters wording:

TURN RAPID SHUTDOWN SWITCH
TO THE "OFF" POSITION TO
SHUT DOWN PV SYSTEM AND
REDUCE SHOCK HAZARD IN ARRAY

Remaining characters:

Capitalized black 3/16 inch minimum height

Section 690.12 rapid shutdown 69

Remaining characters background color:

White

The area on the sign where it says "SOLAR ELECTRIC PV PANELS" in the image is intended to be black letters on a yellow background. **The reason it says PV PANELS rather than PV MODULES** is because only solar professionals call solar modules "solar modules" while most people, including firefighters, refer to solar modules as solar panels.

690.56(C)(1)(b) conductors leaving array level shutdown

690.56(C)(1)(b) refers to rapid shutdown of conductors leaving the array. Within the array there can still be up to 600V on one- and two-family dwellings and up to 1000V on other buildings.

There are specific requirements for a sign that includes specific colors and words. As noted earlier, the NEC does not have the colors in the example of the sign and neither does this book.

The scheme for the 690.56(C)(1)(b) conductors leaving the array level rapid shutdown sign is:

Title wording:

SOLAR PV SYSTEM IS EQUIPPED WITH RAPID SHUTDOWN

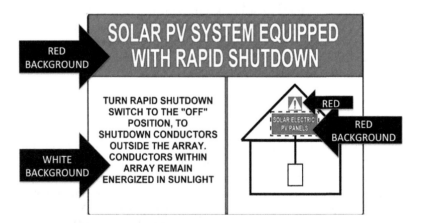

Figure 3.4 NEC Figure 690.56(C)(1)(b) conductors leaving array level rapid shutdown sign

Courtesy NFPA

Title characters:

> Capitalized black 3/8 inch minimum height

Title background color:

> Red (red is the ANSI danger color)

Remaining characters wording:

> TURN RAPID SHUTDOWN SWITCH
> TO THE "OFF" POSITION TO
> SHUT DOWN CONDUCTORS OUTSIDE
> THE ARRAY. CONDUCTORS IN ARRAY
> REMAIN ENERGIZED IN SUNLIGHT.

Remaining characters:

> Capitalized black 3/16 inch minimum height

Remaining characters background color:

> White

The area on the sign where it says "SOLAR ELECTRIC PV PANELS" in the image is intended to be black letters on a red background. Again, only solar professionals call solar modules solar modules while most people, including firefighters, refer to solar modules as solar panels.

Red indicates danger and there is also a red exclamation mark in a triangle on the sign in a proper example.

Both 690.56(C)(1)(a) and (b) labels

Both 690.56(C)(1)(a) and (b) labels shall be located within **3 feet from the service disconnecting means** to which the PV system is connected. If more than one PV system is connected to one service, the sign must identify all rapid shutdown systems on that service. Some large buildings have multiple utility services (e.g. shopping malls). Only the PV systems connected to a specific service need to be identified at a specific service. Other services with PV systems on the same building are identified at their respective services.

Section 690.12 rapid shutdown 71

690.56(C)(2) buildings with more than one rapid shutdown type

If a building has more than one type of PV system with different rapid shutdown types or a PV system installed without rapid shutdown and another system installed later with a rapid shutdown system, then we need a sign that will indicate the different rapid shutdown type scenarios and where the PV system will remain energized after rapid shutdown is initiated.

Figure 3.5 is an example of a sign to be used for a rectangular shaped building with two different types of rapid shutdown arrays. In this example the **color scheme** is:

- Red with white background: Title box on top
- Yellow: 2017 NEC PV array, 80V or less
- Red: 2014 NEC PV array (over 80V)
- Dotted line: within energized

Your building and PV systems will probably look different than the sign in the image, so this sign is customized to the installation. Your sign should plainly indicate to firefighters which areas of the roof will be energized after rapid shutdown is initiated. **A dotted line shall be around areas that remain energized after the rapid shutdown switch is operated.**

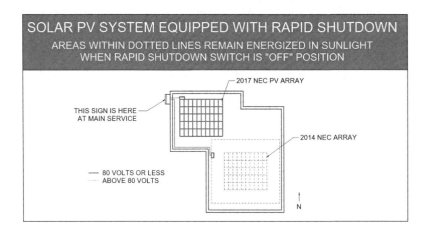

Figure 3.5 Buildings with more than one rapid shutdown type example
Courtesy Robert Price

www.axissolardesign.com

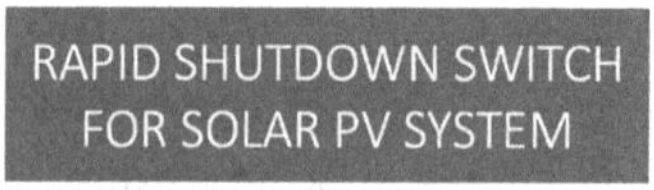

72 Section *690.12 rapid shutdown*

> RAPID SHUTDOWN SWITCH
> FOR SOLAR PV SYSTEM

Figure 3.6 Rapid shutdown sign
Courtesy Sean White

690.56(C)(3) rapid shutdown switch

Within 3 feet of a rapid shutdown switch, there shall be a label with the following words:

RAPID SHUTDOWN SWITCH FOR SOLAR PV SYSTEM

Criteria for rapid shutdown switch label:

- Reflective
- All letters capitalized
- Letters at least 3/8 inch height
- White on red background

2014 NEC rapid shutdown switch label differences:

Since the **2014 NEC** introduced 690.12 Rapid Shutdown of PV Systems on Buildings and 690.56(C) labeling of these systems, the new requirements were purposely left open for interpretation, so that the methods to implement 690.12 and 690.56(C) could be developed with experience over time.

2014 NEC rapid shutdown labeling in 690.56(C) requires similar, but simpler wording than the 2017 NEC:

Permanent plaque or directory including the following wording

PHOTOVOLTAIC SYSTEM EQUIPPED WITH RAPID SHUTDOWN

Plaque shall be:

- Reflective
- Letters capitalized
- Letters minimum height 3/4 inch
- White on red background

The **2014 NEC** had no requirements for the distance of this plaque from the rapid shutdown switch, however, the AHJ is allowed to use common sense and require you to put it near the switch.

There were no other 2014 NEC requirements for rapid shutdown signage, although **2017 NEC signage is a good idea** recommended by the authors of this book in areas enforcing the 2014 NEC.

4 Article 690 part III disconnecting means

Article 690 Part III includes two sections:

690.13 Photovoltaic System Disconnecting Means
690.15 Disconnection of Photovoltaic Equipment

Let us first define **disconnecting means** in general with the **Article 100 definition**:

> Disconnecting Means. A device, or group of devices, or other means by which the conductors of a circuit can be disconnected from their source of supply.

In this definition, we can have a variety of different devices or means that can disconnect sources from supply. These devices are **not required to be load-break rated** devices or big expensive switches that can disconnect a service.

Here are a few examples of disconnecting means:

- Load-break rated dc disconnect
- PV inter-module connectors
- Non-load-break rated disconnect
- Touch safe, tilt-out fuse holders

Even a light switch is a disconnecting means and, for that matter, even unscrewing a light bulb can be considered operating a disconnecting means, no matter how many licensed professional electrical engineers it takes.

Outline of 690.13

690.13 Photovoltaic System Disconnecting Means

690.13(A) Location
690.13(B) Marking
690.13(C) Suitable for Use
690.13(D) Maximum Number of Disconnects
690.13(E) Ratings
690.13(F) Type of Disconnect

> 690.13(F)(1) Simultaneous Disconnection
> 690.13(F)(2) Devices Marked "Line" and "Load"
> 690.13(F)(3) Dc-Rater Enclosed Switches, Open-Type Switches, and Low-Voltage Power Circuit Breakers

690.13 photovoltaic system disconnecting means

Here we are not going to be using a light switch to disconnect an entire PV system; however, we do need to provide a way to disconnect the PV system from all other systems. Recall that the 2017 NEC definition of PV system has been reduced to not include energy storage systems, premises wiring and loads.

690.13(A) location

Disconnecting means shall be installed in a **readily accessible location**.
Article 100 readily accessible definition:
Accessible, Readily (Readily Accessible). Capable of being reached quickly for operation, renewal, or inspections without requiring those to whom ready access is requisite to take actions such as tools (other than keys) to climb over or under to remove obstacles, or to resort to portable ladders and so forth.
Discussion: A readily accessible disconnecting means can be locked in a room or building, but cannot require access with tools. This means that the PV system disconnecting means can be inside a building.
Think of the PV system disconnecting means as the last thing that separates the PV system from any other system.
Here are some examples of common PV system disconnecting means:

- Grid tied inverter backfeed breaker
- Stand-alone system dc disconnect between PV and battery
- Breaker feeding panelboard that is used exclusively for inverters

Recall the disconnecting means in the diagrams in NEC Figure 690.1(b) shown in chapter 1 of this book, beginning on page 10, which **separate the PV system from other systems**.

Figure 4.1 PV system disconnect sign
Image Sean White

690.13(B) marking

690.13(B) Marking requires a sign that says, **"PV SYSTEM DISCONNECT" or equivalent**. It is interesting to note that most label companies go with the "equivalent clause" and add words and make the sign say more words than required, such as "main photovoltaic system dc disconnect."

Other markings are required for some disconnecting means which do not completely deenergize on both sides of the switch when the switch is opened (off). The sign shall say the following or equivalent:

WARNING
ELECTRIC SHOCK HAZARD
TERMINALS ON THE LINE AND LOAD SIDES
MAY BE ENERGIZED IN THE OPEN POSITION

The NEC does not require a particular color or letter size for the PV SYSTEM DISCONNECT sign or the LINE AND LOAD ENERGIZED sign, however, usually this sign is capital white letters on a contrasting background, however, **690.13(B) requires that signs or labels do comply with 110.21(B).**

110.21(B) field-applied hazard markings are required by 690.13(B)

110.21(B) Field-Applied Hazard Markings. Where caution, warning, or danger signs are required by the Code, the labels shall meet the following requirements:

> 110.21(B)(1) The marking shall warn of the hazards using effective words, colors, symbols, or any combination thereof.

110.21(B)(1) *Informational Note* (recommended but not required): **ANSI (American National Standards Institute) ANSI Z535.4–2011. Is recommended.**

> According to this 48 Page **American National Standard for Product Safety Signs and Labels,** there are requirements for the design, use of safety signs and labels.
> Here are some examples:
>
>> DANGER = white triangle, red exclamation mark, red background
>> WARNING = black triangle, orange exclamation mark
>> CAUTION = black triangle, yellow exclamation mark
>> DANGER, WARNING or CAUTION = yellow triangle, black border and exclamation mark.
>
> Since our 690.13(B) sign has the word WARNING in it, we can say that according to ANSI, **WARNING indicates:**
>
>> **"a hazardous situation which, if not avoided, could result in death or serious injury"**

When we have an inverter that has capacitors that discharge within 10 or 30 seconds to comply with rapid shutdown requirements, we still need the 690.13(B) LINE AND LOAD ENERGIZED sign because a switch cover can be opened immediately and be hazardous to an electrical worker that is unaware of the hazard.

690.13(C) suitable for use

Supply side disconnecting means should be listed as suitable for use as a disconnecting means.

If a PV system is connected on the **supply side** of the service disconnecting means (see supply side connection **705.12(A)** page 143) which is **between the meter and the main breaker** in most cases, the disconnecting means shall be **listed as suitable for use as a service disconnecting means.**

Discussion: A **supply side connection** (often and perhaps incorrectly referred to as a **line side tap**) is **not a separate service,** however, it is subject to unprotected currents from the service and it is **now required in the 2017 NEC** that the **supply side connected disconnecting means be listed as being able to be used as a service.** In past versions of the NEC, including the **2014 NEC, this was not a requirement,** but was always **recommended** by solar professionals. As battery systems become more commonplace, these parallel generator connections [230.82(6)] will often be converted into services to accommodate backup loads and EV chargers.

78   *Article 690 part III disconnecting means*

690.13(D) maximum number of disconnects

Each PV system disconnecting means shall not consist of more than six switches mounted in a single enclosure or group of enclosures. It is uncommon to have more than one disconnect for a PV system. An example of where you might have several disconnects to make up a PV system disconnect is with dc coupled systems where more than one dc PV circuit is connected to a battery storage system. Remember that the NEC does not limit the number of PV systems that could be connected to a building [690.4(D)].

A single PV system disconnecting means shall be permitted for the combined ac output of one or more inverters or ac modules in an interactive system. Examples of this may be a microinverter circuit with 15 microinverters or an ac panel main breaker that can disconnect 150 microinverters.

Discussion: If a building has multiple sources of power, then we can have no more than six disconnects per source. This means a building that has one PV system and is connected to the utility can have 12 disconnects.

If the PV system disconnect(s) were grouped together on one side of the building and the utility disconnect(s) were grouped together on the other side, that is not a problem. However, a directory is required at each location to inform about the other sources to the building. Recall that 690.4(D) requires a directory if there are PV system disconnects located in different locations.

690.13(E) ratings

The PV system disconnecting means shall have ratings sufficient for the maximum circuit current, available short-circuit current and voltage that is available at the terminals of the PV system disconnect.

Discussion: We do not just look at the trip rating of an overcurrent protection device. We also look at the maximum available current that can be interrupted. If we are performing a **supply side connection**, we would be on the utility side of all overcurrent protection devices and the overcurrent protection would often need to have a higher **ampere interrupting rating** than a load side connected PV system would. This is one reason why service rated equipment is used for a supply side connection because it often has higher available short-circuit current ratings. The main reason is that service rated switches have a means to connect the neutral to ground in the enclosure.

Overcurrent protection devices have a high-end **ampere interrupting rating** for fault currents and a low-end overload **amp rating**. Often the high-end current is overlooked or ignored, since usually we are focused on the overcurrent protection device protecting conductors with the overload **amp rating**. The overcurrent protection device also needs to protect itself. On your house, the main service disconnect may have a higher **ampere interrupting rating** than the load breakers in the service panel, since it is the first line of defense for utility fault currents. Ampere interrupting ratings are often in the tens of thousands of amps.

690.13(F) type of disconnect

We will discuss the three rules listed as 690.13(F)(1) through (3).

690.13(F)(1) simultaneous disconnection

A **PV system disconnecting means** shall simultaneously disconnect the PV system conductors of the circuit from all other wiring systems.

The PV system disconnecting means shall be an **externally operable** general use switch or circuit breaker **or other approved means**.

A dc PV system disconnecting means shall be **marked for use in PV systems OR** be **suitable for backfeed**.

Discussion: Recall that a PV system disconnecting means is the borderline between the PV system and other wiring systems. This means that a **dc disconnect in an interactive PV system is not a PV system disconnecting means**. It is merely an equipment disconnect for the inverter. Since the interactive inverter is part of the PV system (only processes PV power), the PV system disconnect would be an ac disconnect for an interactive system.

PV system disconnects need to open all conductors simultaneously. A PV module connector is an example of a disconnect or isolating device that does not simultaneously disconnect all conductors. It could not be used as a PV system disconnect, but a connector can be used as an isolation device [690.15].

690.13(F)(2) devices marked "line" and "load"

Devices marked line and load are not permitted for backfeed or reverse current. Most thermally operated circuit breakers cannot tell which way the power is flowing and are suitable for backfeed.

690.13(F)(3) dc-rated enclosed switches, open-type switches, and low-voltage power circuit breakers

Dc-rated, enclosed switches, open-type switches and low-voltage power circuit breakers shall be permitted for backfeed operation.

Discussion: The 2014 NEC used the now-deleted section **690.17 Disconnect Type** to provide a laundry list of disconnects that could be used in a PV system. That list was moved into **690.13 Photovoltaic System Disconnecting Means** and **690.15 Disconnection of Photovoltaic Equipment** and this is where several of the switch options found their new home. The key concept with dc switches is that they can perform differently when current is flowing in one direction vs. the other direction. The products listed in this section are required to be tested in both directions and therefore can be used in PV systems where the current can flow in either direction depending on where the fault occurs.

Outline of 690.15

690.15 Disconnection of Photovoltaic Equipment

    690.15(A) Location
    690.15(B) Interrupting Rating
    690.15(C) Isolating Device

        690.15(C)(1) Connector
        690.15(C)(2) Finger safe fuse holder
        690.15(C)(3) Isolating switch requiring tool
        690.15(C)(4) Isolating device listed for application

    690.15(D) Equipment Disconnecting Means

        690.15(D)(1) Manually operated switch or circuit breaker
        690.15(D)(2) Connector
        690.15(D)(3) Load break fused pull out switch
        690.15(D)(4) Remote-controlled circuit breaker

690.15 disconnection of photovoltaic equipment

First of all, let's discuss the difference between **690.13 PV System Disconnecting Means** and **690.15 Disconnection of Photovoltaic Equipment**.

690.13 PV System Disconnecting Means applies to how to separate a PV system from what is not a PV system. You could also consider this the borderline between the 2017 NEC definition of a PV system and some other system or special equipment covered outside of Article 690.

690.15 Disconnection of Photovoltaic Equipment applies to disconnection of equipment inside the depths of the PV system. Here we can have non-load-break rated equipment, and **non-simultaneous circuit opening, as with module connectors and fuse holders**. We also have equipment disconnecting means, which can be load-break rated and can simultaneously disconnect.

Article 690 part III disconnecting means 81

Although Article 110 does not have a definition for "isolating device" and perhaps it should, there is an **Article 110 definition for isolating switch**, which can give us some clues. Article 100 is titled Definitions.

> **Switch, Isolating.** A switch intended for isolating an electrical circuit from the source of power. **It has no interrupting rating, and is intended to be operated only after the circuit has been opened by some other means.**

The above definition relates to what is often called a **non-load-break disconnect**.

To sum it up, PV equipment disconnecting means are sometimes allowed to be **load-break rated** and other times *not* **required to be load-break rated**.

Another term for load-break rated that we may see on equipment is **current interrupting**.

Some inverters or charge controllers have multiple inputs with a single disconnect; this single disconnect with multiple inputs is acceptable.

The purpose of isolating devices is for the safe ability to work on equipment without being exposed to energized conductors. Isolating devices are not intended to be operated in an emergency condition since they require that the technician makes sure that the circuit being interrupted does not have current flowing. It is essentially a maintenance disconnect.

690.15(A) location

The following are acceptable locations for **isolating devices or equipment disconnecting means**:

- Within equipment
- Within **site and** within **10 feet** of equipment

Equipment disconnecting means **can be operated remote** from equipment if:

- There is a means to **operate the disconnect within 10 feet of equipment**

Discussion: **Equipment disconnects and isolating devices must be within 10 feet of equipment or remotely operated within 10 feet of equipment.**

Many devices, such as inverters and PV modules come with isolating devices or disconnecting means. It is also quite possible that an inverter could be designed with the ability to turn off with the use of a smart phone app. In this case, the smart phone could be used to turn off the inverter and then the connections to the inverter could be unplugged (isolating device) to remove or service the inverter.

690.15(B) interrupting rating

Disconnecting means shall have an interrupting rating sufficient for maximum short-circuit current and voltage.

Isolating devices are not required to have an interrupting rating.

An isolating device is not considered a disconnecting means.

Discussion: PV systems are current-limited and the equipment of a PV system will have short-circuit currents available that are not much more than operating currents. We have to be sure that the terminals of the equipment side of the disconnect can handle these short-circuit currents.

It is the PV system disconnecting means covered by 690.13 that has to have the ability to interrupt currents coming from batteries and the utility, which are outside of the PV system.

Equipment disconnecting means have to be able to open circuits that have current flowing through them.

Isolating devices are not made to interrupt current. Isolating devices should not be opened when there is current flowing.

690.15(C) isolating device

An isolating device is not required to simultaneously disconnect all current-carrying conductors of a circuit.

For example, if someone is removing or installing a PV module, they will not disconnect or connect positive and negative at exactly the same instant. If they were that quick, they would be making millions playing baseball.

Following are the only types of isolating devices we can use.

690.15(C)(1) connector meeting the requirements of 690.33

Must be **listed and identified** for use with specific equipment. The primary purpose for this statement was to make it clear that a connector could be used as a disconnect when the application is clearly defined. For instance, an MC-4 connector is rated up to 50A and can be installed on circuits up to 1000Vdc. However, on a 1000Vdc series string, if you were to take it apart under even a 5A load, it will start a fire. Take that same connector and put it on the input to a microinverter and that same connector can easily break the 5A load without damage to the connector or the operator.

Examples are:

- PV module connectors, such as an MC-4 or Amphenol connector
- Dc-to-dc converter connector
- Some inverters come with ac cable connectors (more often in other countries)

690.33 Connectors are covered on page 101 of this book.

690.15(C)(2) finger safe fuse holder

Finger safe fuse holders are most often used in combiner boxes.

Discussion: When doing voltage or IV curve testing on strings of modules at a dc combiner, we often open the finger safe fuse holders to **isolate the circuit**. This should only be done after the PV output circuit disconnect is opened and there is no current going through the circuit. Many solar installers have witnessed an arc show when opening touch safe fuse holders under load on a system without dc arc-fault protection.

As we can see in Figure 4.2, the finger safe fuse holder can be opened without touching the fuse. After the fuse holder is opened, both sides of the fuse are deenergized.

690.15(C)(3) isolating switch requiring tool

For isolating switches that are accessible to non-qualified personnel, these switches require the use of a tool to move the switch to the open position so that it is impossible for someone without the tool to walk by and open the switch under load. Opening one of these switches under load could be extremely dangerous to the operator.

690.15(C)(4) isolating device listed for application

This option is a catch-all that allows products that have been tested and evaluated for a specific purpose to be used according to how they were evaluated. It may be a combination of several of the options above, but configured in a special way that makes it easier to work on the equipment.

84 *Article 690 part III disconnecting means*

Figure 4.2 Finger safe fuse holder
Courtesy Schurter, Inc.

690.15(D) equipment disconnecting means

Requirements for PV Equipment Disconnecting Means:

- Current interrupting, i.e. load-break rated
- Externally operable
- Will not expose operator to contact with energized parts
- Shall indicate if on or off
- Shall be lockable (in accordance with 110.25)
- Will simultaneously open current-carrying conductors that are **not solidly grounded**

Discussion: It is an important distinction to note in the 2017 NEC, we are required to open all **not solidly grounded** conductors. For inverters that were previously called grounded inverters in the 2014 NEC that were fuse grounded and were the typical US inverter of the

Article 690 part III disconnecting means

2000s decade are now hereby called functionally grounded and are *not* solidly grounded. These formerly "grounded" inverters will have their formerly "grounded conductor" opened in the disconnect in the 2017 NEC. *Not* disconnecting this former "white wire" that was referenced to ground through a fuse was a requirement of the 2014 NEC and the opposite is true in the 2017 NEC.

2014 vs. 2017 NEC inverter grounding

Comparing typical interactive inverter dc grounding
Formerly known as grounded (fuse grounded) inverters:

2014 NEC

- White grounded conductor
- Opened only single ungrounded conductor per circuit
- Fuses when required only on ungrounded conductor

2017 NEC

- Positive and negative must not be white unless one is solidly grounded
- Both positive and negative must be opened at disconnect
- Fuses only required on one polarity (positive or negative)

While we are at it, we might as well give the 2017 and 2014 differences of the formerly known as **ungrounded, transformerless or non-isolated inverters** which are also now known as functional grounded inverters and have the **same rules as all functional grounded inverters in the 2017 NEC.**
Formerly known as ungrounded inverters:

2014 NEC

- Positive and negative must not be white
- Both positive and negative opened at disconnect
- **Fuses required on positive and negative** when fuses required
- **Must use PV wire,** no USE-2 wire for PV circuits outside of conduit

2017 NEC

- Positive and negative must not be white
- Both positive and negative opened at disconnect

86 *Article 690 part III disconnecting means*

> - **Fuses only required on one polarity** (positive or negative)
> - **USE-2 or PV wire both acceptable** for PV circuits outside of conduit
>
> Now that we know about functional grounded inverters, perhaps we can say that the 2014 and earlier versions of the Code were instigating dysfunctional grounding.

690.15(D)(1) manually operable switch or circuit breaker

Examples are:

- A regular circuit breaker
- An ac or dc disconnect

690.15(D)(2) connector meeting the requirements of 690.33(E)(1)

Examples:
   A microinverter or dc-to-dc converter that was listed for load-break disconnection with PV connectors.

   690.33 Connectors covered on page 101.

690.15(D)(3) load break fused pull out switch

Example:
   The same load-break rated fused pull out switch commonly used on air conditioners. These switches can be safely opened under load.

690.15(D)(4) remote-controlled circuit breaker

Example:
   Another term for these circuit breakers is "shunt-trip" circuit breakers. A simple switch closure with no current can signal a very large circuit breaker to open because of the electronics in the circuit breaker.
   End of chapter breakdown:
   Let us break down the different types of disconnecting means as related to PV systems and Article 690 Part III of the NEC.

- **Disconnecting means**
    - Device(s) that disconnect conductors from supply
    - The three below are all disconnecting means.

- **PV system disconnecting means**
  - Separates PV system from *not* a PV system
- **Isolating device**
  - Non-load-break rated
- **PV equipment disconnecting means**
  - Load-break rated

In other countries that do not abide by the NEC, it is common for the only load-break rated disconnect in an interactive PV system to be a backfed circuit breaker in a load center. The rest of the system is connected together with connectors so we can say it is "connectorized." These systems are turned off by the backfed breaker and then taken apart while not under load.

It has been said (although not yet in the NEC) that a well-trained firefighter wielding a fiberglass-handled axe is an exceptional disconnecting means.

5 Article 690 part IV wiring methods

Article 690 Part IV Wiring Methods covers material regarding wiring specific for PV. Much of this material refers to and works with other articles of the NEC, especially the NEC **Articles within Chapter 3 Wiring Methods and Materials**, such as **Article 310 Conductors for General Wiring**.

> 690.3 Other Articles removed from Code in 2017 NEC because the statement was redundant. It was already made clear in 90.3 that Chapters 5–7 modify the first four chapters. Removing 690.3 has nothing to do with trumping or giving up turf. We got rid of it because it was restating the obvious.
> And a reading from the 2014 NEC:
>
>> "**Other Articles.** Whenever the requirements of other articles of this Code and Article 690 differ, the requirements of Article 690 shall apply and, if the system is operated in parallel with a primary source(s) of electricity, the requirements in 705.14, 705.16, 705.32 and 705.143 shall apply."

Although there are PV specific wiring methods in 690 Part IV Wiring Methods, all other parts of the 2017 NEC apply to the wiring of PV systems.

690 Part IV wiring methods articles

690.31 Methods Permitted
690.32 Component Interconnections
690.33 Connectors
690.34 Access to Boxes

Article 690 part IV wiring methods

The majority of 690 Part IV Wiring Methods is contained in 690.31 Methods Permitted.

Outline of 690.31 methods permitted

690.31 Methods Permitted

> 690.31(A) Wiring Systems
> 690.31(B) Identification and Grouping
>
>> 690.31(B)(1) Identification
>> 690.31(B)(2) Grouping
>
> 690.31(C) Single Conductor Cable
>
>> 690.31(C)(1) General
>> 690.31(C)(2) Cable Tray
>
> 690.31(D) Multiconductor Cable
> 690.31(E) Flexible Chords and Cables Connected to Tracking PV Arrays
> 690.31(F) Small Conductor Cables
> 690.31(G) Photovoltaic System Direct Current Circuits on or in a Building
>
>> 690.31(G)(1) Embedded in Building Surfaces
>> 690.31(G)(2) Flexible Wiring Methods
>> 690.31(G)(3) Marking and Labeling Required
>> 690.31(G)(4) Marking and Labeling Methods and Locations
>
> 690.31(H) Flexible, Fine-Stranded Cables
> 690.31(I) Bipolar Photovoltaic Systems

690.31 methods permitted

It is amusing when we see 690 Part IV **Wiring Methods/690.31 Methods Permitted/Wiring Systems**. That is the order of things and how we emphasize the **importance** of our **wiring method systems permitted!** Bill has a proposal to fix this repetition of words in the 2020 NEC.

690.31(A) wiring systems

The following are all permitted wiring methods:

- Raceway wiring methods in the NEC
- Cable wiring methods in the NEC

- Other wiring systems specifically **listed for PV arrays**
- Wiring that is part of a **listed system**

The following are comments on different wiring systems:

- Raceway wiring methods in the NEC
  - Raceway wiring methods are found primarily throughout Chapter 3 of the NEC. Chapter 3 is titled **Wiring Methods and Materials**.
  - EMT is often the raceway method of choice and is covered in **Article 358 Electrical Metallic Tubing: Type EMT**.
- Cable wiring methods in the NEC
  - Cable wiring methods are also found in Chapter 3 of the NEC.
  - USE-2 a popular cable is covered in **Service-Entrance Cable: Types SE and USE. The -2 of USE denotes that it is 90°C rated**.
  - PV wire is another wiring method that was specifically required in the 2014 NEC, but is optional in the 2017 NEC.
- Other wiring systems specifically **listed for PV arrays**
  - A cable manufacturer could design a cable that was superior to PV wire, get a certifying agency to list it, and now we have another cable listed for PV arrays.
- Wiring that is part of a **listed system**
  - While this listed system could be a wiring system for rapid shutdown, it could be any special cable used as part of a listed system. The Engage Cable fits under this category.

Enclosures:
  Where wiring systems with integral enclosures are used, there must be enough extra wire length to facilitate replacement.

300.14 Length of free conductors at outlets, junctions and switch points says that we need **6 inches of free conductor**. It also says, that for boxes less than 8 inches in any dimension, we only need the wire to be able to extend 3 inches outside an opening. Additionally, if the wires are not spliced in the box, then this requirement does not apply.

Table 5.1 Table 690.1(A) correction factors (ambient temperature correction factors for temperatures over 30°C)

Ambient temperature (°C)	Temperature rating of conductor			
	60°C (140°F)	75°C (167°F)	90°C (194°F)	105°C (221°F)
30	1.00	1.00	1.00	1.00
31–35	0.91	0.94	0.96	0.97
36–40	0.82	0.88	0.91	0.93
41–45	0.71	0.82	0.87	0.89

PV source and **PV** output **circuits operating over 30V** in **accessible** locations shall be either:

- Guarded
- In MC cable (Article 330 Metal Clad Cable: Type MC)
- In raceway

For ambient temperatures over 30°C, conductor ampacities shall be corrected in accordance with Table 690.31(A).

Discussion: Table **690.31(A)** Correction Factors is **almost exactly like** Table **310.15(B)(2)(a)** Ambient Temperature Correction Factors Based on 30°C, with a few differences.

- The differences between 690.31(A) and 310.15(B)(2)(a) include:

  - 310.15(B)(2)(a) only has columns for 60°C, 75°C and 90°C rated conductors.
  - Table 690.31(A) has a column for 60°C, 75°C, 90°C and 105°C rated wires.
  - Table 310.15(B)(2)(a) has ratings for temperature increments of 5°C. For instance 30°C, 35°C, 40°C, 45°C, 50°C, 55°C, 60°C.
  - Table 690.31(A) has the same increments except that there are 10°C increments between 61–70°C and 71–80°C. This gives us fewer options with increments in 690, but more increments as far as using 105°C wire in 690.

690.31(B) identification and grouping

PV Source Circuits and PV Output Circuits, unless separated by a partition, **shall not be in the same:**

- Raceway
- Cable tray

Article 690 part IV wiring methods

- Cable
- Outlet box
- Junction box
- Similar fitting

as other conductors from other non-PV systems or inverter output circuits.

If **separated by a partition,** then we have a work around and can put PV source and PV output circuits with other conductors in the same wireway. Partitions are available with some enclosure products such as gutters. Most jurisdictions will require a partition to be part of the listing of the enclosure.

We will also look at identification and grouping of PV conductors.

690.31(B)(1) identification

PV system circuit conductors shall be **identified** at accessible points of

- Termination
- Connection
- Splices
- Exception: evident by spacing or arrangement

Means of identifying PV system circuit conductors:

- Color coding
- Marking tape
- Tagging
- Other approved means
- Spacing or arrangement (690.31(B)(1) Exception)

Only **solidly grounded** conductors in accordance with **690.41(A)(5)** shall be marked in accordance with **200.6 Means of Identifying Grounded Conductors.**

Discussion: **Solidly grounded systems** in the 2017 NEC are rare systems that are *not* **fuse grounded.** An example of a solidly grounded PV system according to the 2017 NEC is a **direct PV well pump,** where the negative conductor is solidly connected to a grounding system that includes a grounding electrode. Grounding through a fuse is not solidly grounding.

These rare and solidly grounded conductors will most likely be marked white according to 200.6 (the marking could be grey, three

Article 690 part IV wiring methods 93

white stripes or three grey stripes). **Fuse grounded current-carrying conductors** that operate at zero volts to ground **are no longer to be considered meeting the requirements of 200.6** and should *not* be identified as white.

The **690.31(B)(1) Identification Exception** tells us that if identification of conductors is evident by spacing or arrangement, then we do not need other forms of identification.

690.31(B)(2) grouping

If the conductors of more than one PV system occupy the same

- **Junction box** or
- **Raceway with removable cover**

then the ac and dc conductors of each system shall be grouped separately by **cable ties** or similar every **6 feet**.

The **690.31(B)(2) Grouping Exception** tells us that the grouping does *not* apply if the circuit **enters from a cable or raceway that is unique and obvious.**

690.31(C)(1) Single conductor cable/general

Single conductor cable types permitted in **exposed outdoor** locations in **PV source circuits:**

- USE-2
- PV wire (listed and labeled)

Different names for PV wire are:

- PV wire
- PV cable
- Photovoltaic wire
- Photovoltaic cable

PV source circuit single conductor cable in outdoor locations must be installed in accordance with **338.10(B)(4)(b)** and **334.30.**

> **338.10(B)(4)(b) is Installation Methods for Branch Circuits and Feeders/Exterior Installations** and tells us that wiring should be installed in accordance with **Part I of Article 225** and should

be supported in accordance with **334.30**. We are also told to comply with **Part II of Article 340**.

Part I of Article 225 refers to:

- Article 225 Outdoor Branch Circuits and Feeders
- **Part I General**

Part II of Article 340 refers to:

- Article 340 Underground Feeder and Branch-Circuit Cable: Type UF
- Part II Installation
  - 340.10 Uses Permitted
  - Uses Not Permitted
  - **Bending Radius [radius of inner edge not less than 5× diameter of cable]**
  - Ampacity [in accordance with 310.15]

334.30 refers to:

- Article 334 Nonmetallic-Sheathed Cable: Types NM, NMC and NMS
- **Section 334.30 Securing and Supporting**

Breaking relevance out of 334.30:

The PV source circuit single conductor cables shall be secured by:

- **Cable ties** listed and identified for securement and support
- Straps
- Hangers
- **Similar fittings designed to not damage cable**

Intervals of support

- 4 ½ feet
- **Within 12 inches of entry into enclosures**

USE-2 and PV wire

USE-2 wire and PV wire are the two most common wiring methods for connecting PV modules to each other and to connecting PV modules to anything else. These wiring methods are commonly installed under PV modules and do get exposed to sunlight. It is interesting that USE stands for Underground Service Entrance and has properties and has been tested for exposure to sunlight. Not usually much sunlight underground.

In previous versions of the NEC, PV wire was specifically required for what was formerly known as "ungrounded" PV arrays, but now there is no specific requirement to use PV wire.

PV wire may be better than USE-2 wire and has been tested with more UV light, but USE-2 wire may be less expensive.

When the PV industry was originally trying to get the code making panel to accept "ungrounded" inverters, they used similar wiring methods as systems with double insulation in Europe.

PV wire is often the wire of choice of PV module manufacturers, since it is acceptable everywhere in the world and with every version of the NEC.

USE-2 wire and PV wire is often colored black and works better when colored black, since black is the color of carbon black, which is a pigment that helps with UV resistance. This is also why black cable ties are more UV resistant. When sourcing black cable ties, use Nylon 6 cable ties when contacting everything except galvanized steel. Nylon 12 is necessary for any cable ties contacting galvanized steel.

There was a time when some inspectors who did not read **200.6 Means of Identifying Grounded Conductors** and they believed that a white USE-2 wire was required.

Some installers have used red USE-2 wire to indicate a positive conductor and as the red faded, people would see white, and think that the formerly red wire was a white grounded conductor.

Most, if not all PV wire is also rated RHW-2 since it has to pass all the same tests as RHW-2. This rating of RHW-2 also allows the conductor to be installed inside buildings. USE-2 if not also rated RHW-2 or XHHW-2 is not permitted to be installed indoors because it may not have been tested for the required fire ratings for indoor wiring.

690.31(C)(2) single conductor cable/cable tray

PV wire with or without a cable tray rating shall be permitted in cable trays outdoors.

PV wire shall be supported in cable trays every 12 inches and secured every 4½ feet.

Article 690 part IV wiring methods

690.31(D) multiconductor cable

Jacketed multiconductor cable assemblies listed and identified for the application are permitted in outdoor locations. This cable shall be secured at least every 6 feet.

Discussion: Microinverter trunk cables are an example of this type of cable.

690.31(E) flexible chords and cables connected to tracking PV arrays

The following shall apply to flexible chords and cables connected to tracking PV arrays:

- Identified as **hard service chord** or **portable power cable**
- Suitable for extra-hard usage
- Listed for outdoor use
- Water resistant
- Sunlight resistant
- Stranded copper permitted to be connected to moving parts of tracking PV arrays in accordance with Table 690.31(E)
- Comply with **Article 400 Flexible Chords and Cables**
- Allowable ampacities in accordance with **400.5 Ampacities for Flexible Chords and Cables**

> Table 690.31(E) Minimum PV wire strands discussion: Tracking arrays have moving parts that may bend wires 365 times per year, which can be tens of thousands of times or more in the life of a 30-year-old PV system.
> Wire with not enough strands will strain-harden and break.

Article **400 Flexible Chords and Cables** Discussion:

> There are different tables to be used for flexible chords and cables, which are used for stationary conductors.

Table 5.2 Table 690.31(E) minimum PV wire strands for moving arrays

PV wire AWG	Minimum strands
18	17
16–10	19
8–4	49
2	130

Courtesy NFPA

The **ampacities in 400.5 and Tables 400.5(A)(1) and 400.5(A)(2) differ from the ampacities typically used for conductors in free air** in Table 310.15(B)(17). These cables also have different ampacities depending on how many conductors are contained within the cable.

The ambient temperature correction factors in 310.15(B)(2)(a) do apply.

690.31(F) small conductor cables

16 American Wire Gauge (AWG) and 18 AWG single-conductor cables are **permitted for module interconnections** if they:

- Meet ampacity requirements of **400.5 Ampacities for Flexible Chords and Cables**
- Comply with correction and adjustment factors from **Section 310.15 Ampacities for Conductors Rated 0–2000 Volts**

Discussion: It may have been unthinkable to use 16 AWG or 18 AWG wire for PV systems when PV was more expensive, however, with the advent of falling prices, it is now thinkable and more likely in large projects using thin film modules with lower than typical crystalline silicon PV module current ratings.

690.31(G) photovoltaic system direct current circuits on or in a building

PV dc circuits inside a building shall be inside of:

- Metal raceway (like EMT Article 358)
- MC cable (Article 330, Metal Clad Cable: Type MC)
- Metal enclosures

This "inside of metal" requirement is **only applied up to the first readily accessible dc disconnect**. After the **first readily accessible dc disconnect**, the conductors are no longer required to be inside metal and normal building wiring requirements apply, along with the wiring methods of 690.31(G)(1) through (4).

The first readily accessible dc disconnect (disconnecting means or disco) shall comply with:

- 690.31(B) Identification and Grouping pages 92–93.
- 690.31(C) Single Conductor Cable page 95.
- 690.15(A) Location [within sight and within 10 feet] page 81.

Discussion: It is possible to have PV dc circuits in NMC (romex) inside a building if it is after the first readily accessible dc disconnect and if the building inspector permits. It is not highly recommended, although it does not violate the NEC.

690.31(G)(1) embedded in building surfaces

If PV circuits are embedded in roofing materials that are

- Built-up roof
- Laminate roof
- Membrane roof

which are typical commercial low-slope roofing materials, **then** the location of circuits shall be clearly marked using a marking protocol that is approved as being suitable for continuous exposure to

- Sunlight
- Weather

Discussion: This requirement is put in place for the protection of roofers, tradespersons and first responders about the location of energized conductors. When someone is working on a roof, they may not be inclined to activate the rapid shutdown system.

This is *not* a requirement for **typical** residential roofs. This provision was a result of practices common with membrane roof attached building-integrated systems (similar to peel and stick products). Installers would slit the insulation below the membrane and insert conduit beneath the membrane roof. This makes for a very clean installation, but it complicates safety in that conduit could be running under access pathways and places a firefighter might use to ventilate the roof. The marking requirement was developed to identify where the conduit was concealed. This marking was also intended to help roofers who may be called upon to cut out roofing and repair the roof. If a roofer is unaware of the existence of a concealed energized conduit, they could easily be injured trying to repair the roof.

Reminder: only applies to dc conductors.

690.31(G)(2) flexible wiring methods

690.31(G)(2) Applies to:

- Flexible Metal Conduit (FMC) smaller than 3/4 inch
- MC cable smaller than 1 inch

Article 690 part IV wiring methods

Either FMC smaller than 3/4 inch or MC cable smaller than 1 inch shall be **protected by substantial guard strips**.

When FMC smaller than 3/4 inch or MC cable smaller than 1 inch is run exposed more than 6 feet from connection to equipment *either* of the following shall apply:

- Follow the building surface
- Protected by approved means

Typically, runner boards are an approved means used to protect flexible wiring methods.

Reminder: 690.31(G) only applies to dc conductors.

690.31(G)(3) marking and labeling required

Warning: photovoltaic power source

This wording above **shall be marked** on the following wiring methods and enclosures that contain PV system **dc** circuits (**not ac!**):

1. Exposed raceways, cable trays and other wiring methods
2. Covers or enclosures of pull and junction boxes
3. Conduit bodies if there are unused conduit openings on the conduit bodies

The marking shall be permanent labels or other approved permanent markings and follow 690.31(G)(4).

Many installers and inspectors have thought that this requirement applies to ac circuits; **this is not a requirement for ac circuits**.

690.31(G)(4) MARKING AND LABELING METHODS AND LOCATIONS

The labeling **locations** for the **WARNING: PHOTOVOLTAIC POWER SOURCE** mark or label shall be:

- Every 10 feet
- Every section separated by
    - Enclosures
    - Walls
    - Partitions
    - Ceilings
    - Floors

Article 690 part IV wiring methods

The **label specifications** shall be:

- Reflective
- Capitalized
- 3/8-inch height minimum
- White letters
- Red background

Label shall be suitable for the environment used.

690.31(H) flexible, fine-stranded cables

Flexible, fine-stranded cables shall be terminated only with terminals, lugs, devices or connectors in accordance with 110.14.

Discussion: Flexible, fine-stranded cables are easy to bend, but are more difficult to terminate. Inexperienced installers in the past have used typical screw terminals that are not meant for flexible, fine-stranded cables. In these cases, the connection will often become loose, resistance will rise and heat will be generated, which can be a fire hazard.

Perhaps 690.31(H) should be an informational note, since the NEC already has provisions for using proper terminations for the application in **110.14 Electrical Connections,** which is where 110.14 is directed.

PV installers in the past used batteries and liked to use flexible, fine-stranded cables, since it would not overstress battery terminals. Most battery cables now use fine-stranded cables with the proper terminations. Batteries are no longer part of a PV system as of the 2017 NEC.

PV arrays that track the sun do require flexible cables and installers of these cables should be aware of the requirements for using proper terminal procedures and equipment.

690.31(I) bipolar photovoltaic systems

Discussion: Bipolar PV systems have a benefit of being able to be considered the voltage for the purposes of the Code as being voltage to ground rather than the maximum voltage between any two conductors.

Analogy:

Your house in the US is wired at 120/240V split phase. This means that you will not have more than 120V to ground in your house, however, you can have an electric dryer that gets all of the benefits of 240V. This is because your house is bipolar in a way and is grounded and has a grounded conductor right in the middle of that 240V.

A bipolar system has two **monopole subarrays** (typically groups of strings), one which is positively grounded and the other negatively grounded. This type of system can give the array half the voltage drop than it otherwise would have if it is configured properly.

Article 690 part IV wiring methods

The wiring rules with these systems state that we cannot have conductors next to each other that could have voltages greater than the wiring method, conductors and Code are rated for in the same location, unless it is inside of an inverter or piece of equipment that can handle these bipolar hot-to-hot voltages that are double of what voltage to ground can be.

In other words, going bipolar does not give a designer the right to have two wires next to each other that have voltages greater than that for which they are rated.

After this **690.31 methods permitted** marathon comes the brief ...

690.32 component interconnections

This section is specific to building-integrated systems.

Fittings and connectors that are **concealed** at the time of **on-site assembly** and **listed** for such use shall be:

- Permitted for **interconnection of modules**
- Permitted for **interconnection of array components**

Fittings and connectors shall be **at least equal to wiring method** in:

- **Insulation**
- **Temperature** rise
- **Fault** current withstand

These fittings and connectors shall also be able to **withstand the environment** in which they are used.

To repeat, section **690.32 Component Interconnections** was designed specifically for building-integrated PV systems.

690.33 connectors

Connectors other than those covered by 690.32 Component Interconnections go here. These connectors are the primary connecters known to solar installers, including your typical MC4 connector or your microinverter cable.

These connectors **shall comply with all of 690.33(A) through (E)**

690.33(A) configuration

Connectors shall be

- **Polarized** (Example: positive or negative, but not both)
- **Noninterchangeable** (with other receptacles of other systems)

Article 690 part IV wiring methods

690.33(B) guarding

Connectors shall **guard persons** against inadvertent contact with

- Live parts

690.33(C) type

Connectors shall be **latching or locking**
  If readily accessible and over 30Vdc or 15Vac

- Then shall **require tool for opening.**

Discussion: PV connectors usually do require a tool for opening, as we will see in 690.33(E)(2) because they are usually not rated for interrupting current.

690.33(D) grounding member

Grounding member should be first to make and last to break.
  Analogy:
If you look at a three-pronged plug in your house, you will see that the grounding member is the longest prong on the plug. This makes it the first to make contact and the last to break contact when plugging in and pulling the plug.

690.33(E) interruption of circuit

Connectors shall be either one of the two following:

1. Rated for interrupting current
2. Require a tool for opening and marked either
    a. Do Not Disconnect Under Load
    b. Not For Current Interrupting

690.34 access to boxes

Junction, pull and outlet boxes (used for wiring) located behind PV modules shall be installed, so that removing the module can make the wiring accessible.

6 Article 690 part V grounding and bonding

Bonding is electrically connecting metal together and grounding is connecting to earth. The term grounding is used often to indicate bonding, as with an equipment grounding conductor, which is used for both bonding metal together and connecting the metal to ground.

690.41 system grounding

System grounding is connecting a current-carrying conductor to ground potential at one place in a system. Not all systems have system grounding, but all metal equipment that could be exposed to a fault will have equipment grounding (AKA bonding).

Outline of 690.41

690.41 System Grounding

    690.41(A) PV System Grounding Configurations

        690.41(A)(1) 2-wire PV arrays with one functional grounded conductor
        690.41(A)(2) Bipolar PV arrays with a functional ground center tap
        690.41(A)(3) Arrays not isolated from the grounded inverter output circuit
        690.41(A)(4) Ungrounded PV arrays
        690.41(A)(5) Solidly grounded PV arrays
        690.41(A)(6) PV systems using other listed and approved methods

    690.41(B) Ground Fault Protection
    690.41(B) Ground Fault Protection Exception

Article 690 part V grounding and bonding

690.41(B)(1) Ground Fault Detection
690.41(B)(2) Isolating Faulted Circuits

690.41(B)(2)(1) Current carrying conductors auto-disconnect
690.41(B)(2)(2) Power off, isolate circuits in functional ground system

690.41 system grounding (continued)

Traditionally system grounding is done by connecting a conductor directly to earth. The grounded conductor is generally colored white for grounded systems. However, new in the 2017 NEC, when system grounding is not solid, such as through a fuse, we no longer color the grounded conductor white. Since the conductor can be at high voltages during a fault, coloring it white can be misleading to an electrician.

690.41(A) PV system grounding configurations

Most of the six configurations are unusual and not worth paying much attention to for most solar professionals. Paying close attention to 690.41(A)(3) is a good idea, since **over 90% of the inverters installed these days are of the non-isolated type.**

690.41(A)(1) 2-wire PV arrays with one functional grounded conductor

This is our old-style fuse grounded solar inverter and was **formerly known as a grounded inverter when applying the 2014 NEC** and earlier. This type of inverter is still common in the large-scale MW inverter realm and seen often when doing maintenance on an old array. This was the most popular inverter in the US in the 2000s.

The term often used for this type of inverter's ground fault detection system is **ground fault detection and interruption (GFDI)**. This inverter bonds a grounded conductor to the inverter internal grounding busbar through an overcurrent protection device (fuse). When there is a ground fault on the ungrounded conductor, currents will flow through the grounding pathway to the GFDI fuse and open the fuse. The inverter will detect that the grounded conductor is no longer grounded and proceed to turn off the inverter and disconnect positive and negative at the inverter in order to prevent fires (you can still get shocked).

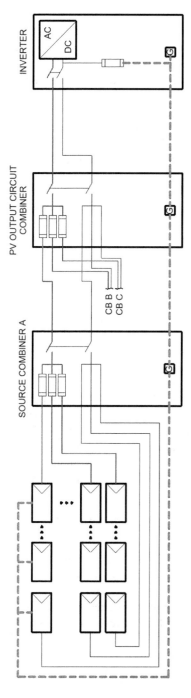

Figure 6.1 Fuse grounded PV array with one functional grounded conductor
Courtesy: Robert Price (2017; modified from Bill Brooks, 2014)
www.axissolardesign.com

One of the reasons that this type of inverter has gone out of style is that there is a blind spot for detecting ground faults. With all PV modules and conductors, things, such as glass and conductor insulation are not perfect insulators and some current will leak through the glass and insulation and complete the circuit through the GFDI fuse. On a small residential system, there is never enough leakage current to cause a problem, but on larger systems, the milliamps add up and the current can be great enough to require the inverter to use a larger GFDI fuse. As the fuse gets larger for these leakage currents, it also inhibits the inverter's ability to detect ground faults. Once there is a ground fault that is unseen (blind spot) then another ground fault can short-circuit the whole PV array and fires have occurred. There are ways that these older systems can be modified by monitoring the current to make them safer, but the easiest thing to do is install a newer non-isolated inverter which is cheaper, safer and more efficient.

690.41(A)(2) bipolar PV arrays with a functional ground center tap

We have mentioned bipolar arrays a number of times in this book and you can look in the index to read about bipolar arrays at every mention. Bipolar arrays are usually only seen in large utility scale PV systems.

690.41(A)(2) applies to these bipolar arrays with a functional ground reference (center tap).

A few people think of bipolar arrays as a way to sneak around the requirement for maximum voltage and double the system voltage. Bipolar arrays will give you double voltage benefits, so there is truth; however, now that we have ground mounted PV arrays of 1500V to ground, there is less of an incentive to double the voltage, since 1500V is a lot and jumping to 3000V for efficiency gains may not be worth doing.

690.41(A)(3) arrays not isolated from the grounded inverter output circuit

This is the most popular inverter today, representing over 90% of inverters installed. **Formerly and commonly known (somewhat inaccurately) as an "ungrounded inverter"** in the 2014 NEC, this inverter is now called a **functional ground inverter** and according to 690.41(A)(3) text, we can officially call this inverter a **not isolated** inverter, or as we prefer a **non-isolated inverter**.

In Figure 6.3 Vin would be input voltage to the inverter. Vs is the source voltage (center-tap split-phase ac transformer).

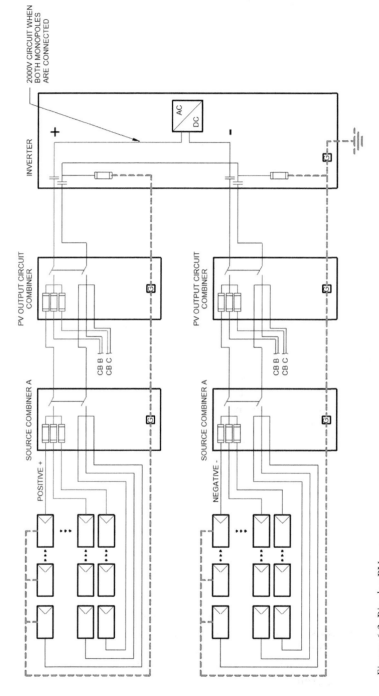

Figure 6.2 Bipolar PV array

Courtesy: Robert Price (2017; modified from Bill Brooks, 2014)
www.axissolardesign.com

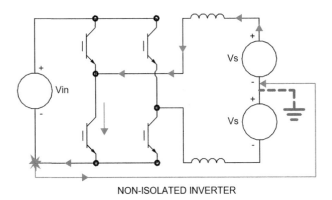

Figure 6.3 Non-isolated inverter showing ground fault pathway
Courtesy: Robert Price (2017; modified from Bill Brooks, 2014)
www.axissolardesign.com

The **non-isolated** gets its name because the inverter is not isolated from the grounded ac transformer. The reason that it was formerly thought of as ungrounded is because the PV array is ungrounded when the system is offline and grounded when operating. When this inverter is operating at 400V, we could expect the positive conductor to measure 200V to ground and the negative to measure –200V to ground. This array is referenced to ground and does not have a voltage that is randomly floating around or isolated by a transformer when the inverter is operating.

These inverters are:

- Cheaper
- Safer
- More efficient

Then your old style, 690.41(A)(1) 2-wire PV arrays with one functional grounded conductor, inverters.
Cheaper because no transformer to manufacture.
Safer because more sensitive ground fault protection.
More efficient since no transformer losses.
Non-isolated inverter ground fault detection:
Non-isolated inverters can detect ground faults much more sensitively than fuse grounded inverters.

Fuse grounded inverters have to have an allowance for leakage currents and the fuse must be upsized as more PV is added to the system.

Non-isolated inverters have different methods of detecting dc ground faults, such as:

- Insulation testing
  - As the inverter wakes up in the morning, a quick pulse of voltage will be sent out along the current-carrying conductors into the array. If there is lower than expected resistance to ground, then there is a failed insulation test and signs of a ground fault. The inverter will not be allowed to start.
- Comparing positive and negative currents (residual current monitor)
  - While operating, if positive does not equal negative, then the electrons must be taking an alternate path, otherwise known as a ground fault.

> Using non-isolated inverters prevents fires since it can be up to 3000 times more sensitive to ground faults than old-school fuse grounded inverters!

690.41(A)(4) ungrounded PV arrays

The 2017 NEC definition of an ungrounded array is different from the 2014 and earlier editions of the NEC. This truly ungrounded PV array is not the *"now known as non-isolated array" (formerly known as ungrounded)*, but is an array that has no functional reference to ground.

Now the only way to have an ungrounded array is with a transformer, but with no fuse or connection between a current-carrying conductor and ground. This array would be considered floating, since the voltage could change with reference to ground. This inverter is uncommon and only used with large utility scale inverters. The array is monitored with a sensitive insulation tester on a continuous basis. This is a very simple and effective method of ground-fault detection.

There is no way to look at an inverter and know whether it is an ungrounded inverter unless you look at the listing label. We are unaware of any ungrounded inverters currently listed in the United States. All the units currently running in the US are in large-scale plants and have been certified to IEC standards.

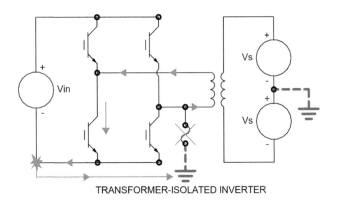

Figure 6.4 2017 NEC ungrounded PV array AKA transformer-isolated inverter
Courtesy: Robert Price (2017; modified from Bill Brooks, 2014)
www.axissolardesign.com

690.41(A)(5) solidly grounded PV arrays

A solidly grounded PV array is one where there is a connection between a current-carrying conductor and ground that is not a fuse and is solid!

Many old style stand-alone systems and some more modern stand-alone systems are solidly grounded. Many dc PV systems are also solidly grounded.

Coming up soon in a 690.41(B) exception, we will mention that if an array is solidly grounded, has one or two PV source circuits and is not on a building, then ground fault protection is not required.

As we can see in Figure 6.5, the grounding busbar is connected solidly to a grounded conductor. This is also an example of a dc direct-coupled PV system, which is common for water pumping systems in the California foothills.

690.41(A)(6) PV systems using other listed and approved methods

PV technology is fast moving and the Code has left open a way for newer inventions to be Code compliant if they are listed and approved. Over the years, several inverters have been certified to this option. One key example in recent history was the Advanced Energy Inverters.

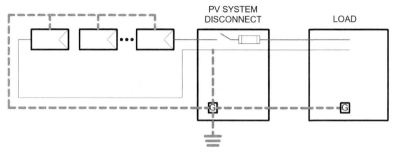

Figure 6.5 Solidly grounded PV array

Courtesy: Robert Price (2017; modified from Bill Brooks, 2014)

www.axissolardesign.com

These larger inverters were configured similar to bipolar PV systems but had several features that were very different from bipolar arrays. Those inverters were among the safest inverters while they were manufactured.

690.41(B) Ground Fault Protection (Direct Current requirement)

Dc PV arrays shall be provided ground fault protection in accordance with **both 690.41(B)(1) Ground Fault Detection** and 690.41(B)(2) **Isolating Faulted Circuits.**

690.41(B) ground fault protection exception

A PV array is not required to have ground fault protection if it meets all of the following requirements:

- One or two PV source circuits
- Not in or on building
- Solidly grounded

We are used to solidly grounded systems with typical electrical systems. The reason it is uncommon with PV systems is because of the lack of high short circuit currents, due to the current-limited characteristics of PV.

Article 690 part V grounding and bonding

690.41(B)(1) ground fault detection

Ground fault detection must meet all of the following requirements:

- Detect ground faults in PV array dc current-carrying conductors
- Detect ground faults in PV array components
- Detect ground faults in functional grounded conductors
- Listed for PV ground fault protection

690.41(B)(2) isolating faulted circuits

Faulted circuits shall be isolated by **either one** of the following of 690.41(B)(2)(1) or 690.41(B)(2)(2):

690.41(B)(2)(1) Current carrying conductors auto-disconnect
Current carrying faulted conductors of the circuit will automatically disconnect.

690.41(B)(2)(2) Power off, isolate circuits in functional ground system
Inverter or charge controller fed by faulted circuit will **automatically do both**

- Stop supplying power to output circuits
- Isolate PV system dc circuits from ground reference in functional ground system

Discussion: A properly UL 1741 listed inverter will have been tested and listed for ground fault protection.

Equipment certification, UL and ANSI

Equipment certification is a de facto requirement of the NEC because it requires all equipment to be evaluated for the intended use in Section 90.7 Examination of Equipment for Safety. Underwriter's Laboratories (UL) has been tasked by the ANSI (American National Standards Institute) to establish certification standards for the safety of PV equipment. There are numerous UL standards that are used for PV equipment. UL1741 is the standard used for certifying inverters.

690.42 point of system grounding connection

Dc PV arrays that are not solidly grounded shall have any current-carrying conductor to ground connection (or reference) made by a ground fault protection device.

For solidly grounded systems, there may only be a single point of system grounding.

Discussion: Ground fault protection is, in essence, determining if there is a connection to ground by a current-carrying conductor that was not meant to be. In order to determine if there is an extra connection (ground fault) then a ground fault protection device must be keeping tabs on the state of the current-carrying conductor vs. ground.

In any electrical system, more than one point of system grounding is called a ground fault. If there were two points of system grounding, then there would be a parallel pathway for current to flow along the grounding system or through equipment.

690.43 equipment grounding and bonding

Exposed metal parts that do not intentionally carry current need to be grounded.

> Outline of 690.43
>
> 690.43 Equipment Grounding and Bonding
>
>> 690.43(A) Photovoltaic Module Mounting Systems and Devices
>> 690.43(B) Equipment Secured to Grounded Metal Supports
>> 690.43(C) With Circuit Conductors

Grounding shall comply with **250.134 Equipment Fastened in Place or Connected by Permanent Wiring Methods (Fixed)-Grounding** and **250.136(A) Equipment Considered Grounded/Equipment Secured to Metal Supports**.

Equipment grounding of PV systems is theoretically the same as equipment grounding of other systems; hence we refer to and take the rules from Article 250 when applying the Code to our PV system grounding systems.

For instance, 690.43 directs us to 250.134, which directs us to 250.32, which directs us to 250.118 and so on and so forth.

Since equipment grounding of a PV system is similar to and takes much of its requirements from Article 250 Bonding and Grounding, inspectors who are not as familiar with PV systems as they are with other systems will often put a lot of effort into inspecting what they know, which is how to inspect grounding and bonding.

> **Why equipment grounding?**
>
> A good reason for equipment grounding is to protect people. If a hot wire got loose and touched a piece of metal that was not grounded, then someone could touch that piece of metal and get shocked. Another good reason for grounding is a pathway for ground fault protection to signal the inverter when there is a ground fault. Without equipment grounding, ground fault protection would not work.

690.43(A) photovoltaic module mounting systems and devices

Equipment that is listed, labeled and identified for bonding PV modules is permitted for equipment grounding and bonding of PV modules. Also, this equipment may be used to bond adjacent modules to each other.
Discussion:

UL 2703 Standard:

Standard for Mounting Systems, Mounting Devices, Clamping/Retention Devices, and Ground Lugs for Use with Flat-Plate Photovoltaic Modules and Panels

When we think of UL 2703 listed racking systems, we often think of a clamping device that holds a PV module or piece of equipment in place and has a sharp point that will pierce the anodized coating of a piece of aluminum and make a Code compliant equipment grounding pathway.

A big benefit to PV installers for using UL 2703 listed racking is the reduction in time and materials required by not having to undertake time consuming grounding methods of the past, such as using lugs on each PV module with a collection of washers, antioxidant chemicals and using emery cloth to rub off the anodized coating of a module frame to get a good connection.

690.43(B) equipment secured to grounded metal supports

Grounded metal supports are support structures that PV modules are mounted on that are already connected to ground by any means

Article 690 part V grounding and bonding 115

permitted in Article 250. In order to bond PV equipment to grounded metal supports, the grounding devices must be:

- Listed = On a National Recognized Testing Lab (NRTL) list
- Labeled = Has a label from the NRTL
- Identified = For specific use (no off-label usage)

Metallic support structure sections shall be **either**:

- Connected via identified bonding jumpers
- Connected to equipment grounding conductor (EGC) and support structure shall be identified

690.43(C) *with circuit conductors*

Equipment grounding conductors for array and support structure shall be run together with the circuit conductors by one of the following means:

- In same raceway
- In same cable
- Run with PV array circuit conductors when leaving array

While it is common for the equipment grounding conductor (EGC) to be run in the same raceway as the current-carrying conductors, the NEC allows the EGC to be run next to the conduit for the current-carrying conductors. For instance, a bare 6 AWG copper conductor could be run in the same trench as a PVC conduit with the current-carrying conductors. The benefit of this buried copper conductor might be lower ground resistance for the whole grounding system.

690.45 size of equipment grounding conductors

PV source and output circuit equipment grounding conductors are sized in accordance with Table 250.122 Minimum Size of Equipment Grounding Conductors, which bases the EGC off of the size of the overcurrent protection device.

If there is no overcurrent protection device (such as with one or two strings) then the value calculated in **690.9(B) Overcurrent Device Ratings** for an assumed overcurrent protection device will be used in place of an overcurrent protective device value.

Recall that 690.9(B) (covered on page 54) calculates an overcurrent protection device based on either

- 125% of maximum current (Isc × 1.56 for PV source circuits)
- Maximum current and application of adjustment and correction factors
- Usable electronic overcurrent protection device rating depending if accessible

Equipment grounding conductor size shall not be smaller than 14 AWG.

Discussion: It is interesting to note that **in many cases multiple PV output circuits are not required to have more than a single 14 AWG** equipment grounding conductor when protected in a raceway. Many solar installers will size the equipment grounding conductor to be the same size as the current-carrying conductor for conductors up to 10 AWG. This is to decrease the possibility of an inspector questioning a 14 AWG equipment grounding conductor.

Just to make a point here, it would be Code compliant to have 8 PV source circuits come off a rooftop in a raceway with 16 different 10 AWG current-carrying conductors along with a single 14 AWG equipment grounding conductor.

Since the smallest equipment grounding conductor is a 14 AWG, which is sized based on a 15A overcurrent device, then we can calculate the highest short circuit current value that will fit on a 15A fuse and use a 14 AWG equipment grounding conductor as follows:

Isc × 1.56 = 15A
Isc = 15A / 1.56 = 9.6A

Therefore, if a PV module has an Isc value of over 9.6A, then it would need greater than a 15A fuse and greater than a 14 AWG equipment grounding conductor.

Table 6.1 NEC Table 250.122 EGC based on OCPD

Rating or setting of automatic overcurrent device in circuit ahead of equipment, conduit, etc., not exceeding (Amperes)	Size (AWG or kcmil)	
	Copper	Aluminum or copper-clad aluminum
15	14	12
20	12	10
60	10	8
100	8	6

Courtesy NFPA

> **Exposed to physical damage = 6 AWG minimum**
>
> Often times the equipment grounding conductors under the array are considered to be **exposed to physical damage** and if that is the case, the bare copper equipment grounding conductor would need to be a **6 AWG** size minimum. Often times in places with significant weather, such as snow and ice, the inspectors will consider under the array to be exposed to physical damage and can require a 6 AWG bare copper equipment grounding conductor.

690.47 grounding electrode system

Are you one of the top PV experts in the country? Lace up your boxing gloves and welcome to 690.47, the section of 690 where there are heated debates on how to interpret how grounding electrode systems are meant to be interpreted in the Code. Try not to take it personally if you have a different idea than your colleagues. Good debate can be fun!

There are many factors to deal with here and grounding electrode systems outside of PV are also hard to understand and are implemented differently regionally. Factors such as wet earth, dry earth, lightning, corrosion, living on a rock, stray currents, proximity to large power production systems and more will influence peoples' opinions on how to properly connect to earth.

Outline of 690.47

690.47 Grounding Electrode System

    690.47(A) Alternating-Current Systems
    690.47(B) Direct-Current Systems
    690.47(C) Systems with Alternating-Current and Direct-Current Grounding Requirements

        690.47(C)(1) Separate Direct-Current Grounding Electrode System Bonded to the Alternating-Current Grounding Electrode System
        690.47(C)(2) Common Direct-Current and Alternating-Current Grounding Electrode
        690.47(C)(3) Combined Direct-Current Grounding Electrode Conductor and Alternating-Current Equipment Grounding Conductor
        690.47(D)(1) Additional Auxiliary Electrodes for Array Grounding

118 *Article 690 part V grounding and bonding*

A grounding electrode system includes pieces of metal connected to earth and other pieces of metal connecting those pieces of metal together and then to the grounding system via a grounding electrode conductor (GEC).

Here are some simple Article 100 Definitions to get us grounded on grounding:

Grounding Electrode. A conducting object through which a direct connection to earth is established.

Grounding Electrode Conductor (GEC). A conductor used to connect the system grounded conductor or the equipment to a grounding electrode or to a point on the grounding electrode system.

Equipment Grounding Conductor (EGC). The conductive path(s) that provides a ground-fault current path and connects normally non-current-carrying metal parts of equipment together and to the system grounded conductor or to the grounding electrode conductor or both. (Also performs bonding).

Ground. The earth.

Grounded Conductor. A system or circuit conductor that is intentionally grounded.

Solidly Grounded. Connected to ground without inserting any resistor or impedance device.

Ground Fault. An unintentional, electrically conductive connection between an ungrounded conductor of an electrical circuit and the normally non-current-carrying conductors, metallic enclosures, metallic raceways, metallic equipment, or earth.

Ground Fault Circuit Interrupter (GFCI). A device intended for the protection of personnel that functions to de-energize a circuit or portion thereof within an established period of time when a current to ground exceeds the values established for a Class A device. (Class A will trip between 4 and 6 mA).

690.47(A) buildings or structures supporting a PV array

A building or structure shall have a grounding electrode system installed in accordance with **Part III of Article 250**.

Article 690 part V grounding and bonding 119

Part III of Article 250 is Grounding Electrode System and Grounding Electrode Conductor and is 5 ½ pages of Code and is where we find the rules for grounding electrodes and grounding electrode conductors.

Here are some hot spots in 250 Part III:

Common types of grounding electrodes:

250.52(A)(1) Metal Underground Water Pipe
250.52(A)(2) Metal In-ground Support Structure
250.52(A)(3) Concrete Encased Electrode
250.52(A)(4) Ground Ring
250.52(A)(5) Rod and Pipe Electrodes (ground rods)
250.52(A)(6) Other Listed Electrodes
250.52(A)(7) Plate Electrodes
250.52(A)(8) Other Local Metal Underground Systems or Structures

250.53 Grounding Electrode System Installation
250.54 Auxiliary Grounding Electrodes

Auxiliary Electrodes, referenced in 690.47 are electrodes that are **attached to the equipment grounding system** and do not follow the rules of typical electrodes.

250.64 Grounding Electrode Conductor Installation
Table 250.66 Size of Alternating-Current Grounding Electrode Conductor

The ac grounding electrode conductor is based on the size of the largest ungrounded service entrance conductor.
Often with PV on a building, the ac electrode is a pre-existing condition. Minimum size is 8 AWG and the dc GEC is found in 250.166 (sounds like 250.66).

690.47(A) buildings or structures supporting a
PV array (continued)

PV array equipment grounding conductors shall be connected to the grounding electrode system of the building or structure that is supporting the PV array.

690.47(A) requires that we follow the rules of

- Article 250 Part VII Methods of Equipment Grounding
- 690.43(C) Equipment grounding conductors to be run with circuit conductors.

Article 690 part V grounding and bonding

- 690.45 Size of Equipment Grounding Conductors
  - Mainly refers to Table 250.122 Sizing Equipment Grounding Conductors

Important clause in 690.47(A):

"For PV systems that are not solidly grounded, the equipment grounding conductor for the output of the PV system, connected to associated distribution equipment, shall be **permitted to be the connection to ground for ground-fault protection and equipment grounding of the PV array.**"

Why this statement is important:

This in effect tells us that the PV systems most commonly used, which are now called functional grounded systems in the US, or any system that is not solidly grounded **does *not* require a separate dc grounding electrode conductor** or system, as was thought in previous versions of the Code by many.

This means that for most systems, **we do not need to have a dc grounding electrode conductor** of any type.

It can also be said that if we were **applying the 2014 NEC, we do not need a dc grounding electrode conductor for typical "ungrounded" non-isolated PV systems.**

Dc grounding electrode conductors are just out of style.

However, for **solidly grounded systems**, which are also out of style, the grounded conductor shall be connected to a grounding electrode via a dc grounding electrode conductor sized in accordance with **250.166 Size of Direct-Current Grounding Electrode Conductor**. Often times the dc grounding electrode conductor ends up being 8 AWG copper, but you need to confirm in the NEC.

690.47(A) informational note on functional grounded systems

This informational note clarifies that most inverters installed in the last decade were actually functional grounded systems and the grounding is done through the connection to the ac equipment grounding conductor.

690.47(B) additional auxiliary electrodes for array grounding

In the **2014 NEC** 690.47(D) Additional Auxiliary Electrodes for Array Grounding said that these electrodes **shall** be installed and got a few people irate. This is now a low-key **optional** thing in the **2017 NEC** that someone can do if they want. One of the reasons that this was controversial is because lightning strikes to the earth can send a wave of voltage. If the wave hits an electrode and causes different voltages at different electrodes that are attached through equipment, then the equipment can sizzle and pop.

Now in the 2017 NEC, these auxiliary electrodes are permitted to be installed, so since they are just permitted, we do not need to worry too much about them.

What they are:

> If you have an array on your roof, you are permitted to go from the array straight to a ground rod. This would also go for a ground mount, however, one way you can look at it is if your array is put in concrete or earth with metal, every post is electrically an additional auxiliary electrode.

Code references for installing these Additional Auxiliary Electrodes for Array Grounding:

250.52 Grounding Electrodes
250.54 Auxiliary Grounding Electrodes
250.66 Size of Alternating-Current Grounding Electrode Conductor
250.54 Auxiliary Grounding Electrodes says that these electrodes can be connected to equipment grounding conductors and shall not be required to comply with electrode bonding requirements or resistance to earth requirements.

690.50 equipment bonding jumpers

Equipment bonding jumpers if used shall comply with 250.120(C)

> 250.120 Equipment Grounding Conductor Installation
>
>> 250.120(C) Equipment Grounding Conductors Smaller Than 6 AWG
>>
>>> Equipment grounding conductors smaller than 6 AWG shall be protected from physical damage by being
>>>
>>> - In a raceway
>>> - In cable armor
>>> - Installed within hollow spaces of framing of building
>>> - In hollow spaces of structures

Article 690 part V grounding and bonding

Discussion: Often times we have frames and rails connected together by UL 2703 listed devices and/or manufactured rail structures.

250.120(C) mentions that protected includes "within hollow spaces of structures" as a protected place. It could be interpreted that underneath a PV module in some cases is "within a hollow space of a structure."

Get grounded! At least functionally.

7 Article 690 part VI to the end of 690

690 Part VI Marking
690 Part VII Connection to Other Sources
690 Part VIII Energy Storage Systems
Other Material no longer in 690 as of 2017 NEC

690 part VI marking

Not all of the requirements for marking PV systems reside within Article 690 Part VI. We also included part of Article 690 Part VI within chapter 4, our Section 690.12 Rapid Shutdown chapter.

Here is a list of the common – and some less common – marking requirements, and where to find them in the NEC and this book, that are **not covered in this chapter in detail.**

- 690.13(B) Line and Load Energized in Open Position Sign
  - Page 76 of this book.
- 690.31(G)(3) and (G)(4) Warning: Photovoltaic Power Source Label
  - Page 99 of this book.
- 690.56(C) Rapid Shutdown Labels
  - Page 67 of this book.
- 705.10 Plaque or Directory at Power Source Disconnecting Means
  - Page 142 of this book.
- 705.12(B)(2)(3)(b) Do Not Relocate Overcurrent Device Label (120% rule)
  - Page 153 of this book.

- 705.12(B)(2)(3)(c) Sum of Breakers Cannot Exceed Busbar Label
  - Page 154 of this book.
- 705.12(B)(3) Supplied by Multiple Sources Label
  - Page 157 of this book.

Now we will cover, in detail, the markings required in 690 Part VI with the exception of the markings covered in the Rapid Shutdown chapter earlier in this book, since we covered the rapid shutdown labeling in chapter 3 of this book.

Outline of 690 part VI

690 Part VI Marking
690.51 Modules
690.52 Alternating-Current Photovoltaic Modules
690.53 Direct-Current Photovoltaic Power Source
690.54 Interactive Source of Interconnection
690.55 Photovoltaic Systems Connected to Energy Storage Systems
690.56 Identification of Power Sources

> 690.56(A) Facilities with Stand-Alone Systems
> 690.56(B) Facilities with Utility Services and Photovoltaic Systems
> 690.56(C) Buildings with Rapid Shutdown [See chapter 3 of this book.]

The marking and labeling requirements for PV systems are included in this book. There are also other marking systems required for energy storage systems in Article 706 Energy Storage Systems and in Article 480 Battery Storage Systems.

690.51 modules

Modules come marked with these data that are required in 690.51:

- Polarity of terminals
- Maximum overcurrent device ratings (usually 15A or 20A)
- The following ratings:
  (1) Open-circuit voltage = Voc
  (2) Operating voltage = Vmp (maximum power voltage)
  (3) Maximum permissible system voltage = maximum system voltage (often 1000V)

(4) Operating Current = Imp = current at maximum power
(5) Short-circuit current = Isc
(6) Maximum power = Pmax = power rating of module

Discussion: Sean recently bought 12 used Arco solar modules that were originally installed in California in the 1980s. It is interesting that there were no data written on the modules. These modules would not be Code compliant to install, unless perhaps a label was made for them in the 21st century. They still work great by the way!

690.52 alternating-current photovoltaic modules

The following ratings are **marked on the ac module**:

690.52(1) Nominal operating ac **voltage** [usually 240V, 208V, 277V or 480V]
690.52(2) Nominal operating ac **frequency** [60Hz in the US]
690.52(3) Maximum **ac power** [inverter output power]
690.52(4) Maximum **ac current**
690.52(5) **Maximum overcurrent device** rating for ac module protection [typically circuit breaker]

Discussion: If an ac module had a PV module that would be 100W when tested dc at STC and contained a 300W inverter, then the ac module would be considered to be a 300W module, although it would never put out even 100W.

Recall that an ac module is, in effect, a microinverter and PV module combined together before testing as a complete ac power-producing unit. The dc wiring, however, is considered as part of the listing and is not subject to the requirements of the NEC. That is the primary difference between an ac module and a PV module connected to a microinverter.

690.53 direct-current photovoltaic power source

Dc PV power source disconnecting means (690.15) requires a permanent label containing:

690.53(1) Maximum voltage [cold temperature corrected]
690.53(2) Maximum circuit current [Isc × 1.25 or dc-to-dc converter(s) maximum rated output]
690.53(3) Maximum rated output current of charge controller or dc-to-dc converter

Discussion: This sign is required to be on or adjacent to the disconnecting means. If the means is a plug connector, and it is connected to a PV array, then the sign would be adjacent to where the plug connectors are located on the inverter. We should consider a specific exception to the sign when the disconnecting means is very close to the PV power source (as with a microinverter).

In the 2014 NEC and earlier we were also required to have the rated **maximum power-point current and the rated maximum power-point voltage**. This is not a safety issue and is **no longer a labeling requirement in the 2017 NEC**.

Since 690.53(3) says "**dc-to-dc converter or charge controller**" we can infer that a charge controller is not always a dc-to-dc converter. When a charge controller does convert dc to dc then a charge controller is also a dc-to-dc converter.

There are no color, size or reflective requirements in the NEC for the 690.53 Direct-Current Photovoltaic Power Source sign.

690.54 interactive source of interconnection

The interactive source of interconnection sign shall have the following:

- Ac nominal voltage [usually 240V, 208V, 277V or 480V]
- Ac rated current [usually inverter power/inverter ac voltage]

Note: ac rated output current can be slightly different than power divided by voltage, due to power factor and operating voltage, which is less than nominal. It is best to determine inverter rated current by reading the inverter label, datasheet or installation manual.

690.55 photovoltaic systems connected to energy storage systems

PV system output circuit conductors shall have the polarity marked when connected to energy storage systems.

Positive and negative shall be marked on the conductors coming from PV, so that the energy storage system is not connected backwards, which could cause exciting safety problems.

> ### Common positive/negative mix-up
>
> Be careful when using PV connectors. A negative connector will only connect to a positive connector. When you are connecting

> a wire to the negative connector of a module, you have to use a positive marked connector, which will not be positive since it is an extension of the negative output of a PV module. Many times, installers get this backwards and are grateful when inverters have reverse polarity protection. It is best to always test voltages before connecting PV to an inverter, charge controller or energy storage system.

690.56 identification of power sources

With PV systems, energy storage systems and the grid, we can have power pushing in all directions, therefore it is prudent to identify what is going on for those who will be puzzled in the future.

690.56(A) facilities with stand-alone systems

Stand-alone (off-grid) PV systems shall have a permanent plaque or directory on the exterior of the building or structure that is in a readily visible location.

The plaque or directory shall indicate that there is a stand-alone system and the location of the stand-alone system disconnecting means.

690.56(B) facilities with utility services and photovoltaic systems

Plaques or directories shall be installed in accordance with 705.10 when PV systems are installed at facilities with utility services.

When there are different power sources at a facility, there shall be a plaque or directory indicating each location for each source of power. This means at the utility service and the PV system disconnecting means, there shall be plaques or directories indicating the different locations of equipment, so that all systems on or in the building may be properly turned off in an emergency situation.

690.56(C) buildings with rapid shutdown (see chapter 3 of this book)

The many required signs for a rapid shutdown system are covered in chapter 3 of this book. Including signs for module level and array level rapid shutdown requirements. See the back cover of this book for proper colors of these signs. Since the NEC is black and white, it can only describe color.

128   *Article 690 part VI to end of 690*

Part VII connection to other sources

Part VII of 690 is short, contains one section and just sends us to another article.

690.59 Connection to Other Sources

Shall be installed in accordance with **Article 705 Parts I and II,** which are:

- 705 Part I General (most of 705)
- 705 Part II Interactive Inverters (1 page of 705)

705 Part III Generators and Part IV Microgrid Systems are a small part of Article 705. The majority of Article 705, including the interconnection requirements in 705.12 are in Part I of 705.

Part VIII energy storage systems

There are two sections in Part VIII. One tells us to go to the new Energy Storage Systems Article 706 and the other is regarding an obscure PV battery combination that has no charge controller.

Outline of 690 part VII

690.71 General
690.72 Self-regulated PV Charge Control

690.72(1) Matching PV to battery
690.72(2) No greater than 3% per hour

690.71 general

Energy Storage Systems connected to PV shall be **installed in accordance with** the new to the 2017 NEC **Article 706 Energy Storage Systems**.

Energy Storage has been taken out of Article 690 in the 2017 NEC and is no longer considered part of a PV system, since they are a separate energy storage system, except for a little bit in 690.72.

690.72 self-regulated PV charge control

Self-regulated PV systems have the benefit of not needing a charge controller, but have the drawback of not being able to work at the optimal part of the IV curve. These systems are often designed so that they will operate towards the Voc end of the IV curve, so that an overcharge will not occur, or they can be designed so that the PV is undersized relative to the battery, yet so is the load. With the advent of less expensive and more reliable PV electronics, self-regulating systems are less common. An example often used for a self-regulating system is a buoy in the ocean with 35 solar cells, charging a 12V battery to operate a flashing light that takes very little energy. Another example is my small solar module that keeps my car battery topped off when I go away for extended periods of time. No charge controller is needed.

690.72(1) *matching PV to battery*

The PV circuit voltage and current rating should be matched to the battery. The NEC does not say what it means by matching, but we can infer that we would like to have a situation where a charged battery would not become an overcharged battery. This could mean that as the battery has more voltage it also gets closer to being fully charged, and that the PV naturally reduces current as it moves towards Voc and away from the maximum power point. We would not want to have a situation where too much current would overcharge a battery in a self-regulating system. A small solar module can keep a larger battery topped off without overcharging, since the small module cannot offer enough current to overcharge.

690.72(2) *no greater than 3% per hour*

As long as we do not charge the battery more than 3% of the rated battery capacity per hour, we will prevent overcharging the battery. The NEC is concerned with dangerous situations and overcharging a battery could cause hydrogen gas and an explosive situation. Also, lead-acid batteries tend to increase their self-discharge rate near full charge, so this helps consume some of the excess current that may be generated near open circuit voltage of the array.

Battery capacity is measured in ampere-hours (Ah).

The NEC has been concerned with lead-acid batteries for years and the new battery technologies that we are seeing often now may not be safe for self-regulating charge control. We do not recommend self-regulating charge control when using lithium batteries on your new Boeing 787 (best planes).

8 Article 691 large-scale photovoltaic (PV) electric power production facility

Article 691 is new in the 2017 NEC and there is not a whole lot to it (about one page in the NEC). There are certain things that can be done with these larger systems, under supervision of an engineer, that cannot be done with smaller systems.

In the past, out excuse for changing the rules on these large "utility scale" solar farms was that we decided to call them utilities and utilities are not subject to the requirements of the NEC. Utilities typically use the National Electric Safety Code (NESC) for designing their systems. The question always has been: Is it really a utility? Now we can use the NEC and no longer have to look for a utility loophole when installing large PV. Also, many jurisdictions did not buy that a large-scale PV system was like utility-owned properties and would enforce Article 690.

Outline of Article 691

691 Large Scale PV
691.1 Scope (>5MW ac)
691.1 Informational Note 1
691.1 Informational Note 2
691.2 Definitions
691.4 Special Requirements

> 691.4(1) Qualified Personnel
> 691.4(2) Restricted Access
> 691.4(3) Medium or High Voltage Connection
> 691.4(4) Loads Only for PV Equipment
> 691.4(5) Not Installed on Buildings

691.5 Equipment Approval

> 691.5(1) Listing and Labeling
> 691.5(2) Field Labeling
> 691.5(3) Engineering Review

Article 691 large-scale PV facilities 131

691.6 Engineered Design
691.7 Conformance of Construction to Engineered Design
691.8 Direct Current Operating Voltage
691.9 Disconnection of Photovoltaic Equipment
691.10 Arc-Fault Mitigation
691.11 Fence Grounding

691 large scale PV

A 5+MW PV system can be installed outside of the scope of 691, for instance on a building. Just because a PV system is larger than 5MW does not automatically make it compliant with 691. It has to meet all of 691.4 Qualified Personnel (pages 23 and 132) requirements as well.

691.1 scope (>5MW ac)

Requirements for PV systems covered in Article 691

- Generating capacity (ac) of no less than 5000kW (5MW)
- Not under exclusive utility control

Generating capacity definition will be covered in 691.2 Definitions, but for the most part is considered the ac output of an inverter. Often times the dc PV portion of a PV system is sized between 1.2 and 1.5 times greater than the inverter generating capacity for large-scale PV projects.

If the PV system were under **utility control, then it would not need to comply with the NEC** at all.

691.1 informational note 1

Large-scale PV covered in this article is producing electricity for the **sole purpose of supplying electricity to a regulated utility.**

691.1 informational note 2

Section 90.2(B)(5) includes information about utility-owned properties not covered by the NEC.

There is a reference to the **National Electric Safety Code,** which is also known as **ANSI Standard C2** and is published by IEEE.

- ANSI is the American National Standards Institute.
- IEEE is the Institute of Electrical and Electronics Engineers.

Article 691 large-scale PV facilities

Section 90.2(B)(5) tells us that **installations that are under the exclusive control of a utility are not covered by the NEC.**

691.2 definitions

Electric supply stations

Locations containing

- Generating stations
- Substations
- Includes generator
- Storage battery
- Transformer
- Switchgear areas

Generating capacity

Ac output is measured in kilowatts at 40°C.

Generating station

A generating station is a plant where electric energy is converted from another source.

Typical sources:

- Solar
- Nuclear
- Chemical
- Wind
- Mechanical
- Hydraulic

691.4 special requirements

Large-scale PV electric supply stations shall be accessible only by authorized personnel.

691.4(1) qualified personnel

Electrical circuits and equipment shall be maintained and operated only by qualified personnel.

Discussion: Qualified person as defined in NEC Article 100 Definitions is:

Article 691 large-scale PV facilities 133

"One who has skills and knowledge related to construction and operation of electrical equipment and installations and has received safety training to recognize and avoid the hazards involved."

This means that there is no hard definition of qualified personnel, however, common sense shall prevail. Some people consider NABCEP Certification as a way of showing that someone is qualified and others believe that an electrician is qualified personnel. Perhaps only people reading this book are truly qualified.

It is interesting that 691.4(1) Informational note directs us to **NFPA 70E Standard for Electrical Safety in the Workplace** for the qualified personnel definition and it is the same definition that we find in NFPA 70, the NEC.

691.4(2) restricted access

Access to PV electric supply stations shall be restricted by fencing or other means in accordance with 110.31. Additionally, field-applied hazard markings shall be applied in accordance with 110.21(B).

110.31 Enclosure for Electrical Installations is in Article 110 Requirements for Electrical Installations **Part III Over 1000V**, Nominal and this section of Code goes into the details of how to keep unqualified people out of the electrical installation.

Some of the other means besides fencing for keeping the PV area restricted to qualified people according to 110.31 are:

- Vaults
- Rooms
- Closets
- Walls
- Roofs
- Floors
- Doors
- Locks

691.4(3) medium or high voltage connection

The connection between the PV electric supply station and the utility shall be made through a **medium voltage or high voltage method,** such as the following:

- Switchgear
- Substation

134 *Article 691 large-scale PV facilities*

- Switchyard
- Similar method

There are different definitions of medium and high voltage, but a general consensus that medium voltage is over 1000V, so we are typically talking about connections here that are **between 4000 and 500,000V**.

691.4(4) *loads only for PV equipment*

The only loads allowed in a PV electric supply station are those to power the auxiliary equipment used in the process of generating power.

Examples of loads would include monitoring equipment, weather stations, lights, controls and PV maintenance equipment.

691.4(5) *not installed on buildings*

Article 691 will not apply to PV systems installed on buildings. This does not mean that PV systems larger than 5MW cannot be on a building; it means that the special rules in Article 691 will not apply to any PV system installed on a building. Article 690 applies in all cases where Article 691 does not apply for PV and the NEC.

691.5 equipment approval

Equipment shall be approved by one of the following methods

691.5(1) *listing and labeling*

A listed PV module tested to UL 1703 and labeled would be an example.

691.5(2) *field labeling*

Typically field labeling would be having a Nationally Recognized Testing Lab (NRTL), such as UL, TUV or Intertek send someone out to approve of the product and give it the lab's label while it is in the field. These field labels may represent a subset of the tests that are conducted on factory labeled equipment since all tests are not feasible in the field.

691.5(3) *engineering review*

When listing and labeling or field labeling are not available, an engineering review can take place to validate that the equipment is tested

Article 691 large-scale PV facilities 135

to an industry standard or practice. Keep in mind this only available if the other two methods are *not* available. It is up to the AHJ weather to approve of this process.

691.6 engineered design

Documentation stamped by a licensed professional engineer shall be made available at the request of the AHJ.

The engineer shall be independent and retained by the system owner.

Additional stamped engineering reports shall be made available upon request of the AHJ documenting:

- Compliance with Article 690
- Alternative methods used not in compliance with Article 690
- Alternative methods used not in compliance with the NEC
- Compliance with industry practice

Discussion: What this means is that under engineering supervision, we are allowed to stray from Article 690 when we meet the requirements of Article 691. Section 691.6 is a documented version of what is commonly performed during the plan check phase of a construction project. Several examples of typical areas where a large-scale PV system is likely to take exception to Article 690, or the rest of the NEC, are listed in 691.8 through 691.11.

691.7 conformance of construction to engineered design

Documentation that **construction of the project followed** the electrical engineered **design** shall be made available to the AHJ upon request.

Additional licensed professional electrical engineer stamped reports detailing that **construction conforms with the NEC, standards and industry practice** shall be provided to the AHJ upon request.

The engineer shall be independent and retained by the system owner.

Discussion: Section 691.7 is a documented version of what is commonly performed during the field inspection phase of a construction project.

691.8 direct-current operating voltage

Included in the documentation required by 691.6 Engineered Design shall be voltage calculations.

691.9 disconnection of photovoltaic equipment

Isolating devices (non-load-break rated disconnecting means) shall be permitted to be more than 6 feet from equipment.
Requirements if isolating devices more than 6 feet from equipment:

- Written safety procedures ensure only qualified persons service equipment.
- Maintenance conditions ensure only qualified persons service equipment.
- Supervision ensures only qualified persons service equipment.

Informational note: Lockout Tagout procedures are in NFPA 70E Standard for Electrical Safety in the Workplace.

Adherence to **690.12 Rapid Shutdown is not required** for buildings whose sole purpose is to protect PV supply station equipment.

Written standard operating procedures detailing shutdown procedures in case of emergency shall be available on site.

691.10 arc-fault mitigation

If PV system does not comply with 690.11 Dc arc-fault protection, then included in the documentation in 691.6 Engineered Design shall be **fire mitigation plans to address dc arc-faults.**

691.11 fence grounding

Fence grounding requirements and details shall be included in the documentation in **691.6 Engineered Design**.

Article 691 overview

When installing a large-scale PV system in the past, PV companies were often forced to pretend that they were a utility. Using Article 690 for compliance with a 100MW power plant was never the intended use of the article. These large systems are unlike smaller systems, in that they are not accessible to people, as are PV systems on buildings and in back yards. With the ability to do things under engineering supervision that stray from the requirements of Article 690, we are no longer put into a position where we have to pretend that we are a utility and that the system is behind a utility fence.

Utilities may still build and operate PV systems that are not required to be compliant with the NEC, however, it is now Code compliant to build a large PV system that does not comply 100% with Article 690. This flexibility can actually improve operation, maintenance, and safety for these large power plants.

9 Article 705 interconnected electric power production sources

Article 705 was first introduced into the NEC in 1987. As both Articles 690 and 705 developed over time, the requirements in 690.64 were incorporated into Article 705. With the advent of much more distributed generation coming onto the grid from solar PV, wind and other sources, Article 705 has grown in size and importance. Now that all the requirements that were in 690.64 have become established in Article 705, the 2017 NEC simplified the text of all of Part VII, Connection to Other Sources, with a simple reference to Parts I and II of Article 705.

In this chapter, we will not cover everything in as much detail as we did when we covered Articles 690 Solar Photovoltaic (PV) Systems and 691 Large-Scale Photovoltaic (PV) Electric Power Production Facility. Instead we will focus on the most important solar PV related material in Article 705.

Article 705 Interconnected Electric Power Production Sources for our purposes includes the requirements for connecting utility interactive inverters to the grid.

Many of the requirements of Article 705 are satisfied by the listing of the interactive inverters.

Outline of Article 705

705 Part I. General
705.1 Scope
705.2 Definitions
705.3 Other Articles
705.6 Equipment Approval
705.8 System Installation
705.10 Directory
705.12 Point of Connection (half of our 705 focus is here)

705.12(A) Supply Side
705.12(B) Load Side

    705.12(B)(1) Dedicated Overcurrent and Disconnect
    705.12(B)(2) Bus or Conductor Ampere Rating

        705.12(B)(2)(1) Feeders

            705.12(B)(2)(1)(a) Feeder ampacity protection
            705.12(B)(2)(1)(b) Feeder overcurrent device

        705.12(B)(2)(2) Taps
        705.12(B)(2)(3) Busbars

            705.12(B)(2)(3)(a) 100% rule
            705.12(B)(2)(3)(b) 120% rule
            705.12(B)(2)(3)(c) Sum rule
            705.12(B)(2)(3)(d) Center fed 120% rule
            705.12(B)(2)(3)(e) Multi-ampacity busbars with engineering supervision

    705.12(B)(3) Marking
    705.12(B)(4) Suitable for Backfeed
    705.12(B)(5) Fastening

705.14 Output Characteristics
705.16 Interrupting and Short-Circuit Rating
705.20 Disconnecting Means, Sources
705.21 Disconnecting Means, Equipment
705.22 Disconnect Device

    705.22(1) Readily Accessible
    705.22(2) Externally Operable
    705.22(3) Indicates Open or Closed
    705.22(4) Sufficient Current Ratings
    705.22(5) Line and Load When Open Situation
    705.22(6) Simultaneous Disconnecting
    705.22(7) Lockable When Off

705.23 Interactive System Disconnecting Means
705.30 Overcurrent Protection

    705.30(A) Solar Photovoltaic Systems
    705.30(B) Transformers
    705.30(C) Fuel Cell Systems (Article 692)
    705.30(D) Interactive Inverters
    705.30(E) Generators

705.31 Location of Overcurrent Protection (supply side)
705.32 Ground-Fault Protection
705.40 Loss of Primary Source
705.42 Loss of 3-Phase Primary Source
705.50 Grounding (Article 250)

Part II. Interactive Inverters
705.60 Circuit Sizing and Current

    705.60(A) Calculation of Maximum Circuit Current

        705.60(A)(1) Inverter Input Circuit Currents
        705.60(A)(2) Inverter Output Circuit Current

    705.60(B) Ampacity and Overcurrent Device Ratings

705.65 Overcurrent Protection

    705.65(A) Circuits and Equipment
    705.65(A) Exception (1) No external sources
    705.65(A) Exception (1) Short-circuit currents do not exceed ampacity
    705.65(B) Power Transformers
    705.65(C) Conductor Ampacity

705.70 Interactive Inverters Mounted in Not Readily Accessible Locations

    705.70(1) Dc Disconnect in Sight
    705.70(2) Ac Disconnect in Sight
    705.70(3) Additional ac Disconnect
    705.70(4) Plaque

705.80 Utility-Interactive Power Systems Employing Energy Storage
705.82 Hybrid Systems
705.95 Ampacity of Neutral Conductor

    705.95(A) Neutral Conductor for Single Phase, 2-Wire Inverter Output
    705.95(B) Neutral Conductor for Instrumentation, Voltage Detection or Phase Detection

705.100 Unbalanced Interconnections

    705.100(A) Single Phase
    705.100(B) Three Phase

Part III. Generators
705.130 Overcurrent Protection (Article 240)

705.143 Synchronous Generators
Part IV. Microgrid Systems (Direct current microgrids are covered in Article 712, which is a new article in the 2017 NEC)
705.150 System Operation
705.160 Primary Power Source Connection
705.165 Reconnection to Primary Power Source
705.170 Microgrid Interconnection Devices (MID)

705.1 scope

Article 705 covers the installation of **multiple power sources connecting in parallel,** such as from a renewable energy source and the utility. One of the power sources must be a primary power source.

705.2 definitions

Interactive Inverter Output Circuit. An inverter output circuit that is connected to a utility grid or other primary power source.

Microgrid Interconnect Device (MID). A device that will allow separation from the grid, such as anti-islanding, which means separating from the grid when the grid is down. Often with ac coupling of PV systems a battery inverter/charger will perform this function.

Microgrid System. A premises wiring system that can perform all or some of the following:

- Generation
- Energy storage
- Loads

A microgrid also can disconnect and reconnect to the primary power source (grid).

Multimode Inverter. An inverter that has the ability to work with the grid or without the grid. Often called a grid-tied battery-backup inverter or a bimodal inverter. Not called a hybrid inverter. This 705.2 definition is word-for-word the same as the 690.2 definition for multimode inverter.

Power Production Equipment. Generating equipment that produces electricity other than the utility service. Examples of power production equipment are:

- Solar PV systems
- Generators
- Fuel cells

705.3 other articles

The entire NEC applies to Article 705, however, there is emphasis on applying the following articles, which are listed in **Table 705.3**:

> Article 445 Generators
> Article 690 Solar photovoltaic systems
> Article 692 Fuel cell systems
> Article 694 Wind electric systems
> Article 700 Emergency systems
> Article 701 Legally required standby systems
> Article 702 Optional standby systems
> Article 706 Energy storage systems (new article in 2017)
> Article 710 Stand-alone systems (new article in 2017)
> Article 712 Dc microgrids (new article in 2017)

705.6 equipment approval

Equipment shall be **listed and or field labeled for intended use**. We need to use equipment that is approved for interconnection to the grid.

Equipment intended to work in parallel with the grid includes:

- Interactive inverters
- Engine generators with special controls
- Energy storage equipment
- Wind turbines

705.8 system installation

Installations only to be performed by qualified person (see pages 23 and 132 of this book for discussion on qualified person).

705.10 directory

A permanent plaque or directory shall be placed at each power source disconnecting means, including the service.

Discussion: If there is an emergency, the firefighters would like to know the location of every power source disconnecting means, so they can know when they turn everything off. This plaque or directory will show at every location where all of the other locations' disconnecting means are located. Firefighters do not want to think they have turned off the building only to find out that they missed a power source disconnect hidden in the backyard.

Article 705 interconnected power sources 143

705.10 has been referenced earlier in section 690.56(B) and has been an important theme in the NEC throughout the years.

705.12 point of connection (half of our 705 focus is here)

Part of the magic of PV systems is connecting inverters to the grid and sending power backwards.

705.12(A) supply side

Supply side connections are permitted.

A supply side connection is connecting a parallel power source, such as a solar interactive inverter on the **supply side of all overcurrent protection** on a service. **If the inverter is on the load side of any overcurrent protection, then it is not a supply side connection.** Typically supply side connections are **between the main breaker and the meter.** A feed-in tariff PV system would also be connected on the supply side of both the main breaker and the service meter. This type of installation could also be covered by 705.12(A).

The **ratings of the sum of all supply side overcurrent devices cannot exceed the rating of the service.** This means that we do not take loads or the size of the main breaker into consideration when determining the maximum amount of inverter rated power that we can connect with a supply side connection. Typically, with a supply side connection, you can connect as much PV as you will ever need. However, some large commercial facilities may want to install PV systems even larger than their service ratings to zero out their electricity consumption over the year. These large systems typically require a significant service upgrade including a larger service transformer.

If disconnecting means is greater than 10 feet from connection point, see **705.31 Location of Overcurrent Protection (supply side)** (page 160 of this book) for information on cable limiters and current-limited circuit breakers for limiting severe utility short circuit currents.

705.12(B) load side

Most solar installers and electricians prefer a load side connection, since it is easy to turn off the main and safely pop in a solar breaker, just like they pop in a load breaker.

We are **permitted to connect** solar on the load side of **any distribution equipment on the premises.**

Examples of distribution equipment eligible for a load side connection include:

- Panelboards
- Switchgear
- Switchboards

Much of the material in 705.12(B) was introduced in the 2014 NEC and the **load side connection material** was **moved from 705.12(D) in the 2014 NEC to 705.12(B) in the 2017 NEC**. Sections 705.12(B) and (C) were removed between the 2014 and 2017 NEC versions since they were properly addressed with the new supply and load side connection categories.

705.12(B)(1) dedicated overcurrent and disconnect

Each source interconnection of one or more power sources in a system shall be made at a dedicated circuit breaker or fusible disconnecting means.

Discussion: In the past, some have interpreted the NEC to say that each inverter needed a dedicated overcurrent protection device, which was especially inconvenient for microinverter enthusiasts. Multiple inverters can have a single overcurrent device as long as those inverters have been listed and tested to work safely with a single overcurrent device. This would be reflected in the installation instructions.

We cannot, however, have a combination load/interactive inverter on a single circuit breaker. This means that you cannot have a PV system that you plug into an outlet. There have been some good ideas that have been taken off the books because of this rule. This is also for good reason. If a homeowner started plugging in a new PV system every week, before they reached a MW, something would catch fire.

705.12(B)(2) bus or conductor ampere rating

We are about to explain how to do some math regarding how much current we can backfeed on a load side connection. In the 2011 and earlier versions of the NEC, we were taught to use the inverter backfeed breaker size for our calculations. After the 2014 NEC we switched over to use 125% of the inverter current in our calculations. In most cases there will be no difference, but here are a few examples of how 125% of inverter current being used in the calculation can be beneficial:

- Rounding up to the next common breaker size
  - Example: If we have a 3kW/240V inverter operating at 12.5A, we then multiply 12.5A × 1.25 = 15.6A. Since there are no 15.6A breakers, we then round up to a 20A breaker. If we use the 125% of inverter current in our calculation rather than the breaker size, we then get an extra 20A – 15.6A = 4.4A to play with.
- Using a 30A breaker, since a 25A breaker is uncommon
  - Example: If we are using a 4kW inverter at 240V then our inverter current is 4kW/240V = 16.7A and 16.7A × 1.25 = 20.8A and in this case we can round up to a 25A breaker, however, electricians may find it difficult to locate a 25A breaker and often use a 30A breaker. It is acceptable to use a 30A breaker in this case as long as the conductor is large enough to be protected by a 30A breaker and as long as the inverter manufacturer allows a 30A breaker to protect the inverter. In this example 125% of inverter current is 20.8A and the breaker is 30A, so we have 30A – 20.8A = 9.2A more of an allowance by using the 125% of inverter current rather than the breaker size method.
- Having a few microinverters on a circuit
  - **Often times with microinverters we do not have the maximum number of inverters on a circuit. At times when the microinverter circuit has a long way to go to reach the interconnection, the designer will place less microinverters on a circuit to address voltage drop considerations. Other times, we see a few microinverters on a circuit, just because it is what fits on the roof.**
  - Example: If we have three 250W microinverters on a 20A breaker at a house, then inverter current would be calculated 250W/240V = 1.04A and 125% of the current for each inverter would be 1.04A × 1.25 = 1.3A and for three inverters 125% of current is 1.3A × 3 = 3.9A. In this case the benefit of using 125% of inverter current rather than the breaker size is 20A – 3.9A = 16.1A benefit, so in this case it can make a big difference.

125% of inverter current is more difficult to explain, but is worth it, since it allows for more PV to be installed than in previous code cycles.

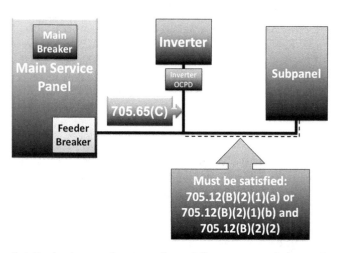

Figure 9.1 Feeder image showing where different parts of the Code apply to different parts of the feeder. Refer back to this image when studying the Codes cited in this image.

Image Sean White

705.12(B)(2)(1) FEEDERS

An example of a feeder is a conductor that is going from a main service panel to a subpanel.

If we are going to connect to the middle of a feeder, we need to make sure that the conductor is properly protected on the load side of the feeder connection. The reason **we are concerned about the load side of the connection to the feeder rather than the supply side,** is because by adding more supply current with our interactive inverter to the feeder, we are no longer protected by the feeder supply breaker. Before the interactive inverter was added, the feeder supply breaker was the protection for the feeder. However, by adding another current source, we can have the potential for overcurrents if we do not comply with 705.12(B)(2)(1)(a) or 705.12(B)(2)(1)(b) which follow.

705.12(B)(2)(1)(a) feeder ampacity protection We add the feeder supply breaker plus 125% of the inverter current for this calculation. If the ampacity of the feeder on the load side of the inverter connection can handle the sum of these currents, then the feeder will be safe. If the existing feeder is not large enough, then we can replace that load side portion of the feeder with a larger feeder. This would rarely ever happen in the field, but the option exists for those that want to use it.

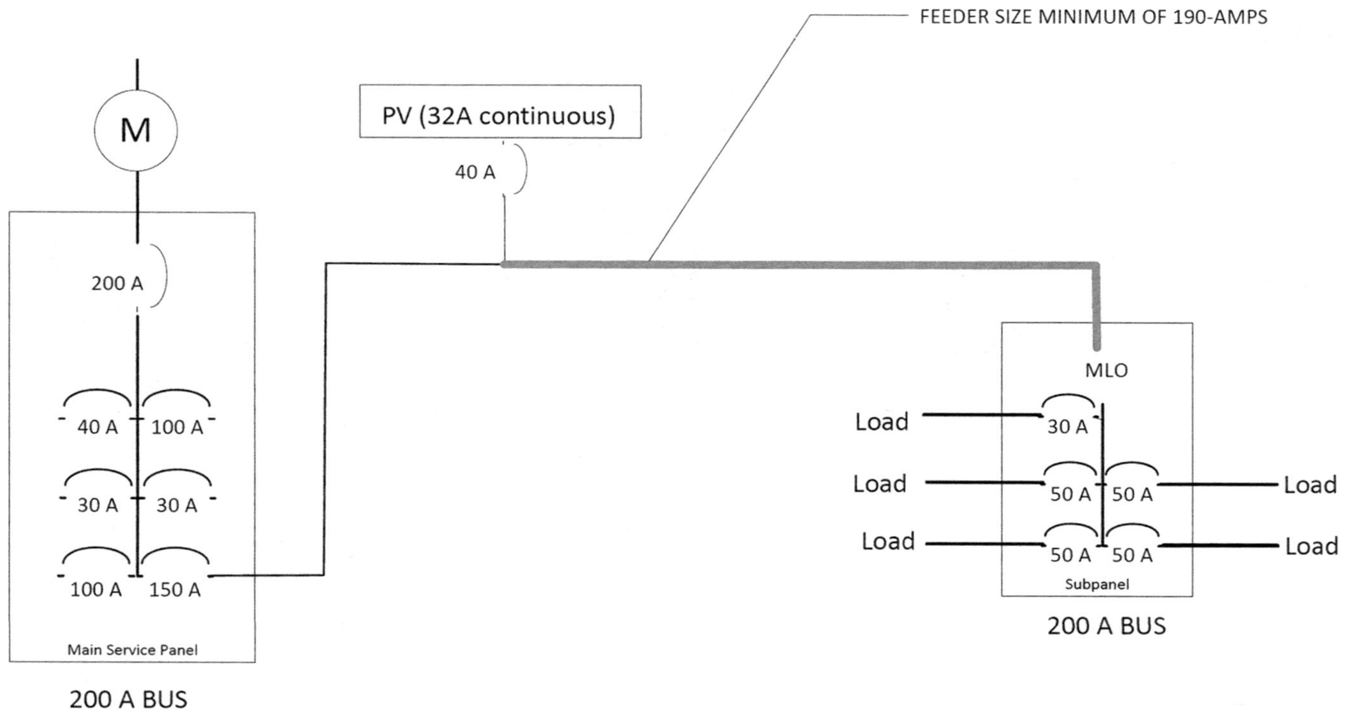

Figure 9.2 705.12(B)(2)(1)(a) sufficient feeder ampacity

Image Bill Brooks

705.12(B)(2)(1)(b) feeder overcurrent device Another option rather than 705.12(B)(2)(1)(a) is to place an overcurrent protection device on the load side of the connection of the interactive inverter to the feeder that is not greater than the ampacity of the feeder.

If we had a 100A feeder, a 100A feeder breaker and a 30A inverter, we could place a 100A breaker on the load side of the connection between the interactive inverter connection to the feeder and the loads.

There have been different interpretations on where this breaker can be placed. The safest place to put the breaker is adjacent to the connection of the inverter circuit to the feeder. Others have the opinion that a "main" breaker in the subpanel will provide this protection. The main reason for putting the breaker adjacent to the PV connection is that no one can argue that taps could be installed between the PV connection and the subpanel that could overcurrent the feeder conductor downstream of the PV connection.

705.12(B)(2)(2) TAPS

When we are connecting an inverter to a feeder, we also need to make sure that the conductor going to the feeder from the inverter is large enough to have a chance to open up the feeder supply breaker in case of a fault (not overcurrent protection, but fault current protection).

Electricians are familiar with the tap rules in 240.21(B) and solar installers are often confused about what a "tap" is since as solar installers we are often and incorrectly calling a supply side connection a "line side tap."

In order to apply the tap rules in 240.21(B), we need an overcurrent protection device protecting the feeder that we are tapping into. A 705.12(A) supply side connection does not have any overcurrent protection on the supply side of the connection, thus cannot follow the tap rules.

The tap rules will add the feeder supply breaker to 125% of the inverter current for this calculation.

There are a number of tap rules, and we will cover the 10-foot tap rule and the 25-foot tap rule for this discussion. Other tap rules will be found in 240.21(B).

10-foot tap rule for solar If the conductor going from the inverter to the connection to the feeder is less than 10 feet, then the ampacity of that inverter output circuit conductor can be no less than 10% of the feeder supply breaker plus 125% of the inverter current.

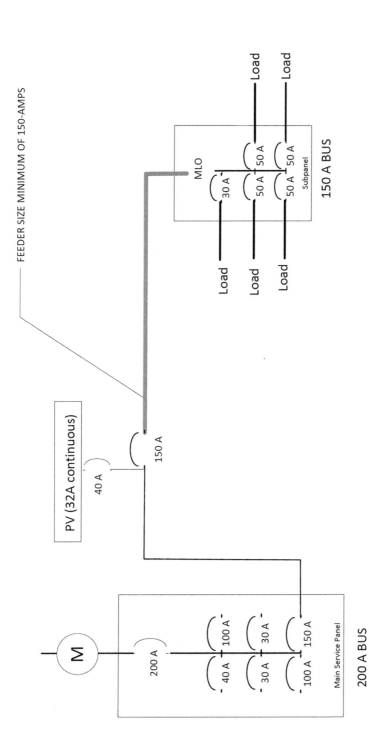

Figure 9.3 705.12(B)(2)(1)(b) Overcurrent device protecting feeder

Note: If 2.5-foot tap rule is applied here, then overcurrent protection can be located up to 2.5 feet from inverter-feeder connection (even at subpanel).

Image Bill Brooks

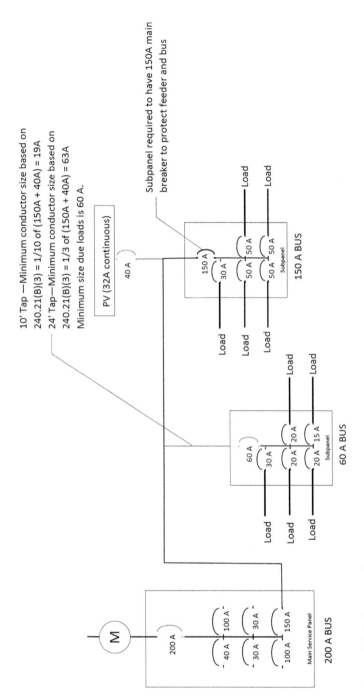

Figure 9.4 Solar tap rules
Image Bill Brooks

Article 705 interconnected power sources 151

Formula:
Inverter output circuit conductor >= 0.1 × ((inverter current × 1.25) + feeder supply breaker)

25-foot tap rule for solar   If the conductor going from the inverter to the connection to the feeder is less than 25 feet, then the ampacity of that inverter output circuit conductor can be no less than one-third of the feeder supply breaker plus 125% of the inverter current.

Inverter output circuit conductor > = 0.33 × ((inverter current × 1.25) + feeder supply breaker)

The reason that the conductor needs to be bigger if the distance is farther is because a long wire has more resistance and is less likely to open up the feeder supply breaker in case of a fault.

Recall that we are sizing the conductor going from the inverter to the connection of the feeder here. We still would need to apply 705.12(B)(2)(1) to make sure that the feeder itself is protected on the load side of the connection point.

The examples for the tap rules are related to PV connections in this section. However, the very same rules are used for existing or new load taps and the required size of conductors for those taps.

705.12(B)(2)(3) BUSBARS

Every solar installer's favorite way to install solar is on a busbar, since popping in a breaker on a busbar is usually the safest and easiest way of installing solar.

Since we have currents coming from different sources on the busbar, we can be creative where we place the solar breakers and often get more out of the busbar than we would think at first glance.

705.12(B)(2)(3)(a) 100% rule   As long as 125% of the inverter current plus the main breaker does not exceed the rating of the busbar, we can place the inverter breaker anywhere on the busbar (it does not have to be on the opposite end from the main supply breaker).

For example, if we have a 225A busbar with a 200A main breaker, we can place a 20A solar breaker anywhere we want on the busbar. We can place up to 25A / 1.25 = 20A inverter anywhere we want on the busbar.

Some installers think that the inverter breaker always has to go on the opposite side of the busbar from the main breaker. This is not true as long as 125% of the inverter current plus the main breaker does not exceed the rating of the busbar.

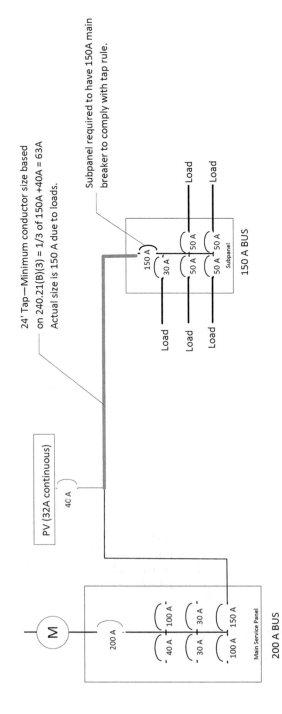

Figure 9.5 25-foot tap rule
Image Bill Brooks

705.12(B)(2)(3)(b) 120% rule

Busbar × 1.2 >= Main Breaker + (1.25 × inverter current)

We can exceed the rating of the busbar by up to 20% after adding the main supply breaker plus 125% of the inverter current as long as the supply breaker and the solar breaker are on opposite ends of the busbar. This opposite end clause was interpreted in such a way that denied center-fed panelboards from being able to apply the 120% rule, but as we will soon see in 705.12(B)(2)(3)(d), this is no longer the case.

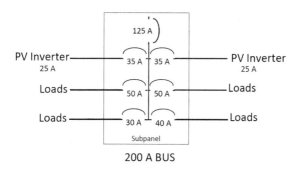

Figure 9.6 100% rule
Image Bill Brooks

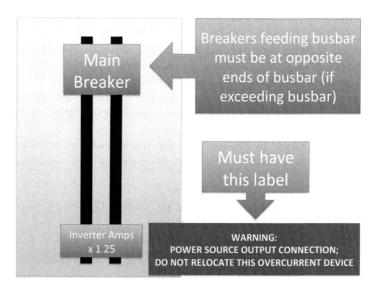

Figure 9.7 705.12(B)(2)(3)(b) 120% rule
Image Sean White

Here is the 120% rule math from a few different angles:

Main + (1.25 × Inv current) <= busbar × 1.2
1.25 × Inv current <= (busbar × 1.2) − main
Inv current <= ((busbar × 1.2) − main) /1.25 or
Inv current <= ((busbar × 1.2) − main) × 0.8 note: (0.8 = 1/1.25)

Maximum inverter power formula using 120% rule:

grid voltage × (((1.2 × busbar) − main) × 0.8) = max inverter power

Recall that backfed breakers will have to be located on the opposite side of the busbar from the main breaker and that there shall be a label saying the following words or equivalent:

WARNING:
POWER SOURCE OUTPUT CONECTION.
DO NOT RELOCATE THIS OVERCURRENT DEVICE.

The reason that we can exceed the busbar rating is because we have currents feeding the busbar coming from different directions, which actually makes it easier on the busbar and prevents busbar "hot-spots" as could happen if the backfeed breaker were put next to the main breaker. The 120% rule will take some heat off of the main breaker when power is fed from a spot on the busbar distant from the main breaker.

705.12(B)(2)(3)(c) sum rule   This rule was primarily created so that subpanels could be logically used for ac combiner panels without the restrictions of the 120% rule. This is a very simple rule that uses the sum of the branch circuit breakers in a panel to protect the busbar of the panel. These branch circuit breakers can be any combination of generation and load breakers.

The easiest way to understand how this rule works is by reading the label required to be installed on the distribution equipment which reads:

WARNING:
THIS EQUIPMENT FED BY MULTIPLE SOURCES.
TOTAL RATING OF ALL OVERCURRENT DEVICES
EXCLUDING MAIN SUPPLY OVERCURRENT DEVICE
SHALL NOT EXCEED AMPACITY OF THE BUSBAR.

In effect, what the "sum rule" is doing is protecting the busbar in reverse, by making sure that the **sum** of all of the breakers on the load side of the busbar protect the busbar.

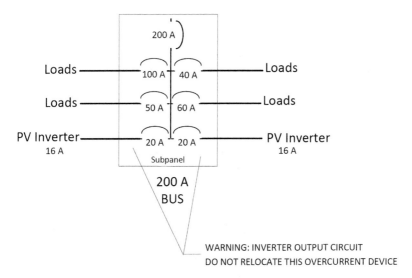

Figure 9.8 120% rule with multiple solar breakers acceptable
Image Bill Brooks

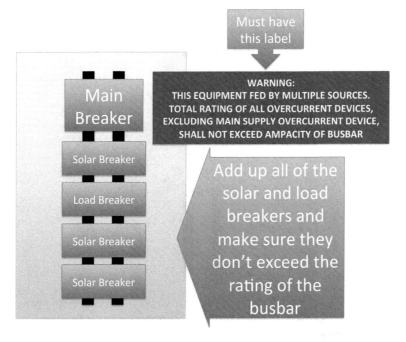

Figure 9.9 705.12(B)(2)(3)(c) sum rule
Image Sean White

If you looked in the main service panel in your house, you would likely find that the sum of the branch circuit breakers is much more than the main breaker of busbar is rated for. We just do not turn on everything at once so the main breaker does not trip.

With the "sum rule" we are protecting the busbar by making sure that we do not have too much current going from the inverter breakers towards the busbar. This rule is conservative if we have loads on the busbar.

One application for someone trying to avoid a supply side connection is to put loads from a main service panel onto a subpanel until you have enough space on your main service panel to apply the "sum rule" and add more solar.

705.12(B)(2)(3)(d) center fed 120% rule   Center-fed panelboards are main service panels or subpanels that are fed *not* from one end, but will have loads connected on both sides of the main breaker.

Millions of dollars were spent on upgrading main service panels, since we could not apply the 120% rule to center-fed panelboards until there was a TIA in the summer of 2016 modifying the 2014 NEC to allow applying the 120% rule to center-fed panelboards. This

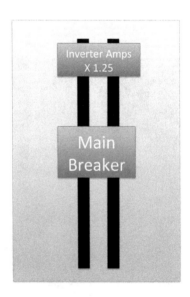

Figure 9.10 705.12(B)(2)(3)(d) center fed 120% rule
Image Sean White

means that if you are using the 2014 NEC or after, you can now apply the 120% rule to center-fed panelboards!

The proper way to apply the **120% rule to center-fed** panelboards is to **only connect solar PV to one side of the busbar and not both.** Perhaps it would be a good idea to make it both sides someday.

705.12(B)(2)(3)(e) multi-ampacity busbars with engineering supervision According to the NEC, there are many things that we can do under **engineering supervision** and connecting to multi-ampacity busbars is one thing that we can accomplish.

In order to make these connections, which would typically be on commercial buildings, we have to have an engineer supervise a study of available fault currents and busbar load calculations.

705.12(B)(3) marking

MARKING SHALL INDICATE PRESENCE OF MULTIPLE SOURCES

Equipment containing a busbar or conductor that is supplied from multiple sources that contains overcurrent devices shall be marked to indicate that it is being supplied by multiple sources.

705.12(B)(4) suitable for backfeed

Backfed breakers shall be suitable for backfeed.

Informational note: Fused disconnects unless otherwise marked are suitable for backfeed.

Discussion: Most circuit breakers are suitable for backfeed. If a circuit breaker is marked line and load, it may not be suitable for backfeed.

Some GFCI breakers are not suitable for backfeed.

Figure 9.11 705.12(B)(3) marking label indicating multiple sources

Courtesy [www.samedaysolarsigns.com/solarsigns/enclosure-contains-conductors-from-multiple-pv-sources get permission]

705.12(B)(5) fastening

Interactive inverter snap-in circuit breaker connections are not required to use an additional fastener to hold them to the busbar.

Discussion: An interactive inverter will immediately stop producing voltage after it is no longer connected to a primary power source (utility). Since we have this anti-islanding capability, we are not in danger of having an energized breaker popping loose and dangling around with voltage with an interactive inverter.

If an inspector requires you to fasten a breaker for an interactive inverter, politely introduce them to 705.12(B)(5).

Authors' note:

Much of the information in Article 705 following 705.12 is also covered more specifically for PV systems in Article 690. For instance, 690.31 covers disconnecting means in more detail than Article 705. Since this book is geared towards solar PV systems, we will decrease the confusion by leaving out redundancies.

Since we will be using UL 1741 listed inverters there are also parts of Article 705 which are satisfied by using listed inverters.

The following are the portions of Article 705 that are not covered in detail in Article 690.

705.14 output characteristics

The output of a power production source shall be compatible with the voltage, wave shape and frequency with which it is connected.

Discussion: Interactive inverters installed in the US must be listed to UL 1741 and will satisfy 705.14.

705.16 interrupting and short-circuit rating

With a PV interactive inverter, the danger of short circuits comes from the utility, rather than the inverter. There are other types of equipment feeding the grid such as generators with rotating machinery that can cause higher fault currents than inverters do.

705.20 disconnecting means, sources

Ungrounded conductors from power production sources must have a means for disconnecting. This applies to most anything in the NEC.

Shop for 2000v breaker on Google

83E8866 - Eaton Cutler-Hammer ...	RD320T32W - Eaton Cutler-...	RXD63S200A - Siemens ...	ITE-Siemens HRD63F200 ...
$40,855.00	**$53,725.00**	**$3,475.95**	**$5,988.47**
Walker Industrial	Walker Industrial	LiveWire Electri...	CapitalElectrica... Free shipping

Figure 9.12 Breakers over 1000V prices
Google Screenshot

705.21 disconnecting means, equipment

We need to have disconnecting means for equipment ungrounded conductors.

If power production source is over 1000V, then disconnecting means are not required. This is because of the difficulty to break an arc at higher voltages and the special requirements for systems over 1000V. Higher voltage systems typically employ large circuit breakers controlled by expensive relaying systems.

705.22 disconnect device

Should be manual or power operated and:

1 Readily accessible **same as 690.13(A)**
2 Externally operable and able to open by hand if power failure
3 Indicates weather on or off
4 Properly rated for available voltage and current **same as 690.13(E)**
5 Have sign as indicated in **690.13(B)** if line and load may be energized in open position.
6 Simultaneous disconnection of all ungrounded conductors **same as 690.13(F)(1)**
7 Lockable in off position **same as 690.15(D)**

160   *Article 705 interconnected power sources*

705.23 interactive system disconnecting means

Readily accessible disconnecting means is required.

705.30 overcurrent protection

Conductors shall be protected from all sources in accordance with **Article 240 Overcurrent Protection.**

705.30(A) solar photovoltaic systems

Solar systems shall be protected in accordance with **Article 690.**
  Section **690.9 is overcurrent protection** for solar PV systems.

705.30(B) transformers

Transformers in accordance with Section 450.3.
  Article 450 is Transformers and Transformer Vaults and 450.3 is Overcurrent protection.
  Consider one side of the transformer and then the other side as the primary. Same as section **690.9(D)**.

705.30(C) fuel cell systems

See Article 692 Fuel Cell Systems.

705.30(D) interactive inverters

In accordance with 705.65.

705.30(E) generators

In accordance with 705.130.

705.31 location of overcurrent protection (supply side)

Important addition to 705.12(A) supply side connection!

Due to excessive currents coming from the utility due to short-circuits, there are two methods of overcurrent protection that shall take place within 10 feet of the connection point of the supply side connection.

- **705.31 Method: Overcurrent protection** within 10 feet of connection point to service entrance conductors (between meter and main breaker).

- 705.31 Exception Method: Cable limiters or **current limited circuit breakers** to be installed on each ungrounded conductor within 10 feet of connection point.

705.32 ground-fault protection

Interactive systems shall be installed on the **supply side of ground-fault protection equipment.**

705.32 exception (connecting to the load side of ground fault protection)

If there is ground fault protection from all ground-fault current sources, then it is acceptable to connect to the load side of ground-fault protection equipment.

Discussion: It is difficult to protect against ground-faults if there are current sources on each side of ground-fault protection. Many solar professionals do not bother trying to connect on the load side of ground-fault protection. All new services of 1000A and larger, built after 2002, are required to include ground-fault protection. Most manufacturers of these systems are accustomed to providing documentation that their detection circuits can handle backfed power from a generator on the load side of the ground-fault protector.

705.40 loss of primary source

We are not allowed to feed the grid, when the grid (primary power source) is down!

Discussion: UL 1741 Inverters are safe with regards to not feeding the grid when the grid is down, AKA, anti-islanding.

Since Article 705 includes other sources of power being connected to the grid, including big diesel generators, there is a provision in 705 to require conductors to be disconnected automatically from the primary power source.

Interactive inverters, however, will stay connected and monitor the grid-connection and export power five minutes after a good clean grid power source is indicated. Interactive inverters cannot feed the grid when the grid is down. No exceptions. It cannot happen and it is certified to prevent such an occurrence.

Multimodal inverters can power loads when the grid is down and this is done with a separate output circuit coming out of the inverter that is not connected to the utility grid. Multimodal inverters have multiple (usually two) outputs that operate in different modes, such as an interactive mode output and a stand-alone mode output.

705.42 loss of 3-phase primary source

When a single phase is down, 3-phase inverters shall stop exporting power.

Interactive inverters do not have to disconnect all phases when a phase is down, since listed interactive inverters will stop exporting power when one phase is down.

Other than interactive inverters must disconnect all phases.

It is not mentioned here in 705.42, but it is acceptable to have single phase inverters connected to a 3-phase service and if a phase is down that is not connected to a single-phase inverter, then the single-phase inverter is allowed to export power on the good phases.

Often single-phase microinverters are connected to 3-phase services via a cable that rotates the phases by connecting different inverters to different phases. See **705.100 Unbalanced Interconnections.**

705.50 grounding (Article 250)

Grounding done in accordance with Article 250 Grounding and Bonding.

Also see 690 Part V Grounding and Bonding which is 690.41 through 690.50 and on page 103 of this book.

Part II interactive inverters

Much of Part II duplicates 690, so we will skip a few repetitions.

705.60 circuit sizing and current

See 690.8 Circuit sizing and current.

705.60(A)(1) inverter input circuit currents

Maximum rated current for an interactive inverter input circuit is the maximum rated input current of the inverter.

It is interesting to note that 705.60(A)(1) contradicts 690.8, since according to 690.8 we would size the input to an interactive inverter based on the PV currents rather than the maximum current an inverter will allow on the input. There are proposals for the 2020 NEC to fix this contradiction and allow what is in Article 705 to be used in Article 690.

705.60(A)(2) inverter output circuit current

Same as 690.8(A)(3) for PV.

Article 705 interconnected power sources

705.60(B) ampacity and overcurrent device ratings

Same as 690.8(B).

705.65 overcurrent protection

Same as 690.9.

705.65(B) power transformers

Same as 690.9(D).

705.65(C) conductor ampacity

Besides sizing the wire based on our standard wire sizing techniques used throughout the Code, including checks for overcurrent protection devices, continuous current and adjustment factors. We also need to check wire size based on the tap rules based on the feeder breaker without adding 125% of the inverter current.

See wire sizing in chapter 12 of this book.

705.70 interactive inverters mounted in not readily accessible locations

If an inverter is located in a location that is *not* readily accessible, such as a roof, then the following requirements shall be met:

- 705.70(1) Dc disconnecting means in sight of inverter
    - Dc disconnect shall be in sight or within the inverter
- 705.70(2) Ac disconnecting means in sight of inverter
    - Ac disconnect shall be in sight or within the inverter
- 705.70(3) Additional ac disconnect
    - In compliance with 705.22, which is also covered in 690.13, which must be readily accessible, visible, lockable, simultaneous disconnection of current-carrying conductors (not a connector).
- 705.70(4) Plaque
    - Plaque in compliance with 705.10, which means there will be a plaque at each power source disconnecting means showing

164 *Article 705 interconnected power sources*

where all of the other power source disconnecting means are located. 705.70(4) requires a plaque and in 705.10 you are also allowed a directory.

This section was directly copied from the disconnecting means section of Article 690 several code cycles ago. Items 705.70 (1) and (2) can be connectors. Item 705.70(3) cannot be a connector in this section. There are many types of directories including a map of the site showing the location of all the disconnects. In addition to a map, a directory can be a text description of the location of each disconnect.

705.80 utility-interactive power systems employing energy storage

Interactive power systems with energy storage must be marked with:

- Maximum operating voltage
- Equalization voltage (if equalization is performed)
- Polarity of grounded conductor

705.82 hybrid systems

Hybrid systems permitted with interactive inverters. Hybrid system example would be adding generator, wind or hydro to a PV system.

705.95 ampacity of neutral conductor

- **705.95(A) Neutral Conductor for Single Phase, 2-Wire Inverter Output**
    - When connecting a single-phase 2-wire inverter to split phase or 3-phase systems, the sum of the inverter current plus the load currents on any phase cannot exceed the ampacity of the neutral.
    - When PV sources are going backwards, currents will do the opposite of what they would do when currents are going to loads and therefore will not cancel out neutral currents from other phases.
    - For example, if you have a 120V inverter on a 120/240V circuit and there are loads on the opposite phase as the inverter, then the neutral will have the sum of the currents. Since loads on a 120/240V circuit cancel out neutral currents, then PV currents will add to the currents on the neutral when on a different phase. This is reverse logic of a load. PV is the anti-load.

- Note: It is best to place an interactive inverter on the circuit with the loads in order to cancel out the load currents.

- **705.95(B) Neutral Conductor for Instrumentation, Voltage Detection or Phase Detection**
  - Most interactive inverters installed in the US do not use the neutral conductor to carry currents, however, UL 1741 requires a neutral to be monitored in most cases. This neutral conductor that is used by the inverter to monitor the grid for problems is not required to be larger than the equipment grounding conductor.

705.100 unbalanced interconnections

- **705.100(A) Single Phase**

  Single-phase inverters (and ac modules) shall be connected to 3-phase systems in order to limit unbalanced voltages to not more than 3%.

  Discussion: Just as loads can be strategically placed to limit unbalances on 3-phase busbars, so can single-phase inverters, but with reverse logic as with loads. If a building is out of balance and **inverters are placed on the phases with lower voltage,** which are the most heavily loaded phases, the inverters will reduce the utility load on those phases and help raise the voltage by sending power in a reverse direction. The result is that the imbalance of the service is reduced and the neutral current also goes down. Inverters cause voltage rise on a backfed busbar.

- **705.100(B) Three Phase**
  - Three-phase inverters (and ac modules) shall have all phases de-energize upon unbalanced voltage in a single phase unless the inverter is designed so that significant unbalances will not result.
  - Discussion: UL 1741 listed inverters will automatically disconnect with large imbalances. There are new inverter standards such as UL 1741 SA, which will change the way inverters help support the utility grid and other standards that may have been released while you were reading this book. These new developing standards address a new level of saturation of distributed generation on the grid and can differ based upon where you live. Exciting times!

Part III generators

In the IEC, a PV system is called a PV generator, but in the NEC we are talking about a spinning **motor that makes electricity**. Some PV systems are made into hybrid systems by adding a generator.

705.130 overcurrent protection

In compliance with **Article 240 Overcurrent Protection** with special attention to protecting from multiple sources.

705.143 synchronous generators

When generators are operated synchronously with the grid, there must be special equipment to ensure synchronicity. Going out of sync can cause damage to equipment.

Part IV microgrid systems

Article 705, Interconnected Power Production Sources is by nature dealing with alternating current, since our electric distribution systems are based on alternating current. The microgrid systems covered here in Article 705 are **ac microgrids**. These microgrids covered in Article 705 have the ability to disconnect from the utility and operate in stand-alone mode. **Many solar professionals call smaller ac microgrids ac coupled systems.**

Dc microgrids are covered in Article 712, which is a new article in the 2017 NEC.

705.150 system operation

Microgrid systems are permitted to disconnect from the primary power source (grid) and operate separately from the grid.

705.160 primary power source connection

Connections to primary power sources (grid) shall comply with **705.12 Point of Connection.**

705.165 reconnection to primary power source

Reconnection shall be provided with necessary equipment to establish a synchronous connection. Interactive and multimode inverters all provide this service.

705.170 microgrid interconnection devices (MID)

See **705.2 Definitions** in NEC or on page 141 of this book for the definition of MID.

UL 1741-listed interactive and multimode inverters provide this service. This is about the ability of a microgrid to automatically anti-island and disconnect from the grid whenever the grid goes down or is not up to the specifications of voltage, current and waveform.

End of chapter thoughts:

> At this point, we have covered the meat and bones of PV and the NEC with Articles 690 and 705. Rather than going **meat and bones level** and creating a book that would be too heavy to carry and cover many things that are not specific to PV systems, we will go **broth level** on the rest of the Code.
>
> Solar planet loving vegetarians, have no worries, the meat and bones used to make this book are actually soy based simulated meat products. Yum!

10 Storage articles

As PV systems grow at an exponential pace and head toward the goal of someday saturating the grid, energy storage systems are a natural fit for bottling sunlight for nighttime photon usage.

This chapter will cover:
Article 480 Storage Batteries
Article 706 Energy Storage Systems
Article 710 Stand-Alone Systems
Article 712 Direct Current Microgrids

With energy storage being in vogue at all of the solar conferences this decade (or millennia), we will take a ride on the **Article 480 Storage Batteries** train and see where that track crosses paths with the new **Article 706 Energy Storage Systems** express at an intersection with the new **Article 710 Stand-Alone Systems** (formerly Section 690.10). An honorable mention will go to another new article, Article **712 Direct Current Microgrids**.

Batteries and energy storage

In the 2014 NEC and earlier, **Article 480 Storage Batteries** was used along with **Section 690.10 Stand-Alone Systems** and **690 Part VIII Storage Batteries** for designing and installing energy storage for stand-alone PV systems. Much of the order has changed in the 2017 NEC, since **energy storage has been taken out of Article 690,** and the new articles mentioned above have been added. Much of the changes are with the organization of the NEC, rather than changing the installation drastically.

When work started on Article 706, the intention was to take the information from Articles 480 and 690 and get rid of those sections.

Somewhere along the way, the lead-acid folks decided they still needed Article 480 even though it is in the wrong chapter of the NEC and is no longer needed. The focus for field installations should be on Article 706 since storage batteries are never used by themselves – they are always part of an energy storage system.

Scopes of Article 480 Storage Batteries and Article 706 Energy Storage Systems:

- Article 480 scope applies to all stationary installations of storage batteries
- Article 706 scope applies to all permanently installed energy storage systems above 50Vac and 60Vdc

As we can imagine, the scopes cross paths and both articles will need to be considered in battery systems with voltages above 50Vac and 60Vdc.

Article 480 Storage Batteries has been in the NEC longer than we have been alive and Article 706 Energy Storage Systems is brand new. As you read 480, you will realize that much of the material is related to good old time tested and reliable lead-acid battery technologies.

As we can see by reading Article 706, it focuses more on new modern technologies including but not limited to chemical batteries. Article 706 will cover the higher voltage lithium battery modules that are being sold, flywheel storage, capacitor energy storage and compressed air storage. Perhaps someday Article 480 will go away as it was originally planned.

Article 706 covers the energy storage "**system**" while Article 480 covers the batteries (really only lead-acid batteries). In general, it is best to simply ignore Article 480 and focus on Article 706.

When designing an energy storage system with storage batteries, we should pay close attention to Article 706. We will notice that there are many reoccurring themes in this article and within the material in Articles 690 and 705 with regards to having separate energy sources with the ability to operate in parallel and the requirements of disconnecting systems from each other.

An outline of noteworthy NEC **energy storage articles, parts and sections follows**. *Italics are author comments.*

Article 480 storage batteries

480.1 scope

Applies to all stationary applications of storage batteries.

170   *Storage articles*

This means that Article 706 does not overrule 480 (another possibility for contradictions in the NEC).

480.7(D) disconnecting means/notification

Apparently, battery systems below 60V are not required to have disconnects. This was a last-minute change to the NEC and should not be followed as it creates very real safety hazards. This is another reason to disregard Article 480 in general at this point.

Marking shall include

1   Nominal battery voltage
2   Maximum available short-circuit current
3   Date short-circuit current calculation performed

   • *Short-circuit information provided by battery suppliers*

4   Marked in accordance with 110.16 Arc-Flash Hazard Warning.

480.10 battery locations

480.10(A) ventilation

Ventilation appropriate to battery technology
   Lead-acid and nickel-cadmium batteries can create explosive hydrogen gasses.

480.10(B) live parts

Live parts shall be guarded in accordance with **Section 110.27 Guarding of Live Parts.**

480.10(C) Spaces about battery systems

Spaces around batteries shall comply with **Section 110.26 Spaces about Electrical Equipment.**

Article 706 energy storage systems

706.1 scope

Article 706 applies to permanently installed energy storage systems over 50Vac or 60Vdc.

706.2 definitions

Energy storage system (ESS) definition

"One or more components assembled together capable of storing energy for use at a future time. ESS(s) can include, but are not limited to **batteries, capacitors, and kinetic energy devices** (e.g., **flywheels and compressed air**). These systems can have **ac or dc output** for utilization and **can include inverters and converters** to change stored energy into electrical energy."

- **Energy Storage System, Self-Contained.** A unit that is factory assembled and shipped as a unit.
- **Energy Storage System, Pre-Engineered of Matched Components.** A system that is field-assembled with specific components that have been engineered to work together as a unit.
- **Energy Storage System, Other.** A system that is neither self-contained nor pre-engineered but are assembled as a system (almost all battery systems used with PV systems to date have been of this variety).

Flow battery definition:

"An energy storage component **similar to a fuel** cell that stores its active materials in the form of **two electrolytes external to the reactor interface.** When in use, the electrolytes are transferred between reactor and storage tanks."

706.3 other articles

Whenever the requirements of other articles of the NEC and Article 706 differ, 706 shall apply.

This "other articles" clause used to be, but is no longer in Article 690 and perhaps it will find its way out of 705 someday.

706.7(D) disconnecting means/notification

Markings required

1. Nominal ESS voltage
2. Maximum available short-circuit current
3. Associated clearing time or arc duration

172 *Storage articles*

4 Date calculation performed

- Battery equipment suppliers can provide short-circuit information.
- NFPA 70E Standards for Electrical Safety in the Workplace has requirements for arc-flash labels.
  - *Unlike PV, which is current-limited, ESSs can have excessive short circuit currents which can be more dangerous with regards to arc-flashes.*

706.30 installation of batteries

706.30(A) dwelling units

ESS for dwelling units shall not exceed 100V to ground

706.30(A) exception

Where live parts are not accessible during routine maintenance, ESS voltage exceeding 100V is permitted

Many of the new ESS sold as units including electronics operate over 100V.

706.30(B) disconnection of series battery circuits

Battery circuits subject to field servicing shall have provisions to disconnect series connected circuits into **240V or less segments.**

Non-load-break disconnects are acceptable.

706.30(C) storage system maintenance disconnecting means

Systems exceeding 100V shall have disconnecting means that disconnect grounded and ungrounded conductors of ESS, that do not disconnect grounded conductor of other electrical systems.

- 706.30(D) Storage System of More Than 100V
  - *Systems greater than 100V to ground are permitted to operate without grounded conductors if ground fault protection is installed.*
- Part IV Flow Battery Energy Storage Systems
  - Flow batteries use pumps to move electrolyte and are new to the NEC in 2017.

- 704.40 General *(Flow Batteries)*
  - *Unless otherwise directed by Article 706,* flow batteries shall comply with the provisions of Article 692 Fuel Cell Systems.

Article 710 stand-alone systems

In the 2014 NEC, this material was covered in 690.10. The 2017 NEC has separated energy storage systems from PV systems.

710.1 scope

"This article covers electric power production sources operating in stand-alone mode."

710.15(A) supply output

Supply shall be at least as much as the largest load.
   Often times off-grid system owners will turn on a generator when using high power loads.

710.15(C) single 120V supply

Often a 120V inverter is used in stand-alone systems and 120/240V designed service equipment **may be used** (bonding L1 to L2).

- *The neutral bus must be rated greater than the sum of the loads.*
- *No multi-wire branch circuits since neutral currents will not cancel each other out and a sign must read:*

WARNING:
SINGLE 120-VOLT SUPPLY. DO NOT CONNECT
MULTIWIRE BRANCH CIRCUITS!

710.15(D) energy storage or backup power requirements

"Energy storage or backup power supplies **not required**"
   Direct PV systems are stand-alone systems, which only work when the sun is shining.

710.15(E) back-fed circuit breakers

Back-fed plug-in type circuit breakers for **stand-alone** *circuits* **must be secured** *so that pulling on breaker will not remove breaker.*
   Interactive circuits are not required to be secured.

Article 712 direct current microgrids

712.2 definitions

Many of these definitions clarify other areas of the Code, such as with types of system grounding.

Direct current microgrid (dc microgrid)

A direct current microgrid is a power distribution system consisting of more than one interconnected dc power source, supplying dc-to-dc converter(s), dc load(s), and/or ac load(s) powered by dc-ac inverter(s). A dc microgrid is typically not directly connected to an ac primary source of electricity, but some dc microgrids interconnect via one or more dc-ac bidirectional converters or dc-ac inverters.

A dc coupled PV system that is grid-interactive and provides power to dc loads fits the definition of a dc microgrid.

Grounded 2-wire dc system

A system that has a solid connection or reference-ground between one of the current-carrying conductors and the equipment grounding system.

Reference grounded dc microgrids are similar to functional grounded PV systems.

Definition could be solidly or not-solidly grounded.

Nominal voltage

A value assigned to a circuit or system for the purpose of conveniently designating its dc voltage class.

- An example is that a 12V lead-acid system will charge over 14V and sit fully charged at 12.6V.
- Nominal is often defined as "in name only."

Reference-grounded system

A system that is not solidly grounded but has a **low-resistance electrical reference to ground** that maintains voltage to ground in normal operation.

A *fuse grounded system is an example of a reference-grounded system, which is not solidly grounded and should **not** have a white-colored grounded conductor as of the 2017 NEC.*

Resistively grounded

A system with a high-resistance connection between the current-carrying conductor and the equipment grounding system.
These are used in industrial grounding systems.

712.30 system voltage

712.30(A) and (B) Solid and reference grounded system voltages are defined as voltage to ground.

712.30(C) Resistive grounded and ungrounded system voltages are between conductors.

712.35 Disconnection of Ungrounded Conductors (Disconnecting Means)

Ungrounded, reference grounded and resistive grounded systems shall open all "non-solidly grounded" current-carrying conductors.
Only solidly grounded conductors are **not** *opened.*

712.37 Directional Current Devices (disconnecting means)

If the device is for single current direction, it must be listed and marked for single current direction.
Examples of single current direction devices are **magnetically quenched contactors** *and* **semiconductor switches.**

712.52(B) Over 300V (Wiring methods/system grounding)

Dc microgrids over 300V shall be **reference grounded** or **resistively grounded.**
Solidly grounded or ungrounded dc microgrid systems over 300V are not allowed.

Article 625 Electric Vehicle Charging Stations contains section **625.48 Interactive Systems** that will direct **electric vehicles that have the ability to backfeed the grid** to Article 705. Section 625.48 also indicates that these **bidirectional batteries on wheels** need to be listed for exporting power.

When you are connecting your electric vehicle to your house as an optional stand-by system, you are encouraged to follow the Code and use Article 702 Optional Standby Systems. Article 702 is most often used when backing up a house with a generator. There are some YouTube videos that show people how to connect electric cars to their houses in order to operate in stand-by mode in a non-Code compliant way. Be careful! Do not have too much fun!

As you and your customers contemplate implementing energy storage into a grid-connected PV project, be sure to understand the regulations and incentives for connecting energy storage to the grid. Policies are evolving and the Public Utilities Commissions are working hard to determine how energy storage will be brought into the grid.

A good source for finding out about incentives for all things renewable is the Database of State Incentives for Renewables and Efficiency at www.dsireusa.org.

11 Chapters 1–4, Chapter 9 tables and Informative Annex C

So far, we have covered the NEC articles that are used often in the solar industry, including 690, 705, 480, 706, 710 and 712. These articles often refer to other articles in the NEC and this chapter will go over the different articles of the NEC from a PV perspective.

First, we will cover Chapters 1 through 4 of the NEC, which apply generally to all electrical installations. We will then mention the relevance and when to use articles in Chapters 5 through 7. Finally, we will cover Chapter 9 and look at the informative annexes that we can use when designing PV systems.

Chapters 1 through 4 have been referenced throughout this book, especially with respect to Article 690 and 705.

Here are some of the top articles to be familiar with in Chapters 1 through 4:

- Article 100 Definitions
- Article 240 Overcurrent Protection
- Article 250 Grounding and Bonding
- Article 310 Conductors for General Wiring [especially the 310.15(B) tables!]
- Article 358 Electrical Metallic Tubing: Type EMT

Chapters 1 through 4 of the NEC

The NEC is divided into Chapters and the number of the chapter precedes the three-digit number of the article. For instance, Chapter 6 Special Occupancies includes Article 690 Solar PV Photovoltaic Systems and Chapter 1 General begins with Article 100 Definitions. Chapters 1 through 4 apply to all electrical installations, including PV systems.

An outline of **often used for PV articles, parts and sections** in Chapters 1–4 of the NEC follows. *Italics are author comments.*

Recall the hierarchy of NEC organization: Chapters/Articles/Parts/Sections.

Chapter 1 general

Article 100 definitions

This is a great resource for defining terms throughout the NEC, including "PV."

Article 110 requirements for general installations

110.21 marking

This is often referred to in 690 and 705.

110.26 spaces about electrical equipment

This is often 3 feet back and 30 inches wide or the width of the equipment for wider equipment.

Table 110.28 enclosure types

This includes NEMA enclosure types for wet areas, wind, dust, etc.

Chapter 2 wiring and protection

Article 200 use and identification of grounded conductors

200.6 means of identifying grounded conductors

Applying 200.6 is less common for 2017 NEC designed PV systems than for 2014 NEC designed PV systems since the new functional grounded inverter definition requires no dc grounded conductor marking. See page 16.

Article 230 services

Supply Side Connections need to be properly bonded using best practices similar to service equipment since they are exposed to utility currents. See 705.12(A) page 143.

Article 240 overcurrent protection

240.4 protection of conductors

240.4(B) OVERCURRENT DEVICES RATED 800 AMPERES OR LESS

You can round up over the ampacity of the conductor to the next higher overcurrent device size.

240.4(D) SMALL CONDUCTORS

"Small conductor rule"

- *12 AWG copper wire needs at least 20A overcurrent protection.*
- *10 AWG copper wire needs at least 30A overcurrent protection.*

240.6 standard ampere ratings
- *Special fuses*
  - *1A, 3A, 6A and 601A*
- *Fuses and breakers*
  - *5A increments 15 to 50A*
  - *10A increments 50 to 110A*
  - *25A increments 125 to 250A*
  - *50A increments 250 to 500A*

240.21(B) FEEDER TAPS

See 705.12(B)(2)(2) page 148.

- 240.21(B)(1) 10-foot tap rule
- 240.21(B)(1) 25-foot tap rule
- 240.21(B)(5) Outside taps of unlimited length

Article 250 grounding and bonding

690 Part V is also Grounding and Bonding; see page 103.
690 Part V often refers to 250.

Chapters 1–4, Chapter 9 tables, Annex C

Part III grounding electrode system and grounding electrode conductor

This is where experts differ and there is confusion.

250.52(A) GROUNDING ELECTRODES PERMITTED

(1) Metal Underground Water Pipe (*common*)
(2) Metal In-ground Support Structure
(3) Concrete-Encased Electrode *(common)*
(4) Ground Ring
(5) Rod and Pipe Electrodes *(common)*
(6) Other Listed Electrodes
(7) Plate Electrodes (*common in Canada*)

250.53 grounding electrode system installation

Learn how to install electrodes here.

Table 250.66 size of ac grounding electrode conductors

The size of the ac grounding electrode conductor (GEC) is based on the size of the largest ungrounded service entrance conductor.

Usually, when dealing with PV on an existing service, you already have an existing ac grounding electrode conductor and do not need to go to **Table 250.66.**

Part VI equipment grounding and equipment grounding conductors

250.122 SIZE OF EQUIPMENT GROUNDING CONDUCTORS (EGC)

EGC sizes are based on the size of the overcurrent protection device. See page 116.

250.166 SIZE OF THE DIRECT-CURRENT GROUNDING ELECTRODE CONDUCTOR

For most PV systems, including functional grounded systems, there are no dc grounding electrode conductor requirements. In these systems, the **ac equipment grounding conductor and ac grounding electrode conductor provide reference and a pathway to ground.**

690.47(A) paragraph 3 "For PV systems that are not solidly grounded, the equipment grounding conductor for the output of the PV system, connected to associated distribution equipment, shall be permitted to be the connection to ground for ground-fault protection and equipment grounding of the PV array."

In past Code cycles, it was common for an AHJ to require a separate dc electrode or dc grounding electrode conductor for most PV systems. It is now clear that this is no longer the case.

An example of a PV system that requires a dc grounding electrode conductor is a dc direct PV water pumping system.

Chapter 3 wiring methods and materials

Article 300 general requirements for wiring methods and materials

300.7 raceways exposed to different temperatures

300.7(A) SEALING

If a raceway is going to areas of different **temperature changes**, seal the raceway at the junction of the temperature change to **prevent condensation** where warm air would otherwise meet cold pipe.

300.7(B) EXPANSION

If a raceway is long enough and exposed to enough changes in temperature, calculations similar to voltage temperature calculations can be made to determine the need for expansion joints or flexible conduit. These calculations also apply to the expansion of solar rails.

Article 310 conductors for general wiring

310.15 ampacities for conductors rated 0–2000V

310.15(A)(2) SELECTION OF AMPACITY

If different sections of a circuit have different ampacities, then the lowest ampacity shall apply unless the shorter section is **less than or equal to 10 feet or 10% of the circuit length**, whichever is less.

310.15(B) TABLES [MOST OFTEN USED PAGES IN THE NEC!]

For examples see chapter 12 "Wire Sizing" of this book.

310.15(B)(2)(a) ambient temperature correction factors based on 30°C   Used to derate conductor ampacity in hot places, such as outside in the summer.

Recommended by authors to use ASHRAE 2% average high temperature data found at www.solarabcs.org/permitting.

Table 690.31(A) shall be used for solar PV wires rather than Table 310.15(B)(2)(a), however, Table 690.31(A) is the same 90+% of the time as Table 310.15(B)(2)(a). See page 91 of this book.

310.15(B)(3)(a) adjustment factors for more than three current-carrying conductors   Four or more conductors in conduit or cable will have trouble dissipating heat and shall be derated.

Not to be applied if distance is less than 24 inches.
Not to be applied to neutrals carrying only unbalanced currents.

Table 310.15(B)(3)(c) Ambient Temperature Adjustment for Raceways or Cables in Sunlight on a Roof (In 2014 NEC and *removed in 2017 NEC*)

Table 310.15(B)(3)(c) is *not in the 2017 NEC!*

310.15(B)(3)(c)(not a table) is in the 2017 NEC and says that if our raceways or cables are **7/8 inch or less above rooftop**, we shall add a 33°C adder.

It is recommended to be at least 1 inch above rooftop to avoid the adder and to keep debris from building up on the rooftop.

310.15(B)(16) Allowable Ampacities of Insulated Conductors (for conductors not in free air)

This is the most used page in the NEC. It gives the ampacities of conductors in cables, raceways, direct buried and everything else that is not in free air.

310.15(B)(17) Allowable Ampacities of Insulated Conductors (for conductors in free air)

Table 310.15(B)(17) covers ampacities of conductors in free air, such as PV wire underneath solar arrays.

Articles 320–362 various cables and conduits

From EMT to NMC (Romex), you will find it here.

Article 352 rigid polyvinyl chloride conduit: type PVC

PVC is most often used in Hawaii due to high corrosion rates.

Article 358 electrical metallic tubing: type EMT

EMT conduit is the most often used wiring method among solar installers on the **mainland United States**.

358.30(A) securely fastened

The EMT is to be secured every 10 feet (3 feet from junction boxes and equipment).

Article 392 cable trays

Cable trays are often used around arrays to manage wires. PV wire is allowed in cable trays.

Chapter 4 equipment for general use

Article 480 storage batteries

Article 490 equipment over 1000V, nominal

490.2 High voltage is defined as 1000V for this article.

Chapter 5 special occupancies

If you are installing a PV system in a barn, on a gas station, an aircraft hangar or other special occupancy, you will need to comply with the special requirements in **Chapter 5** Special Occupancies.

Some examples of special occupancies where you may be installing PV are:

- Hazardous locations
- Commercial garages
- Aircraft hangars
- Gas stations
- Bulk storage plants
- Facilities using flammable liquids
- Health care facilities
- Assembly occupancies (for over 100 people)

- Theaters
- Amusement parks
- Carnivals
- Motion picture studios
- Motion picture projection rooms
- Manufactured buildings
- Agricultural buildings
- Mobile homes
- Recreational vehicles
- Floating buildings
- Marinas

Just remember, if you are installing PV in a special place, such as one of the locations listed above, you should take Chapter 5 into consideration.

Chapter 6 special equipment

Article 690 which is what most of this book covers is the most important article in **Chapter 6 Special Equipment,** and there are more articles that may relate to PV installations that we shall mention, such as:

Article 625 Electric Vehicle Charging System
Article 645 Information Technology Equipment
Article 646 Modular Data Centers
Article 647 Sensitive Electronic Equipment
Article 670 Industrial Machinery
Article 680 Swimming Pools, Fountains and Similar Installations
Article 682 Natural and Artificially Made Bodies of Water
Article 692 Fuel Cell Systems
Article 694 Wind Electric Systems

If your PV system incorporates wind, fuel cells or electric car charging, you should look to other articles in **Chapter 6 Special Equipment.**

Chapter 7 special conditions

There are many other special conditions besides the interconnection to the grid, which we already covered in chapter 9 of this book, where we covered Article 705. We also already covered other special conditions in this book in chapter 10 where we covered energy storage related articles, such as Article 706 Energy Storage Systems, Article 710 Stand-Alone Systems and Article 712 Direct-Current Microgrids.

Other articles in Chapter 7 Special Conditions that we may have to reference one day include:

> Article 701 Legally Required Standby Systems
> Article 702 Optional Standby Systems
> Article 720 Circuits and Equipment Operating at Less Than 50V
> Article 750 Energy Management Systems

Chapter 8 communication systems

Chapter 8 Communication Systems is not subject to the requirements of Chapters 1 through 7 of the NEC except where referenced in Chapter 8. These systems are separate from PV systems, although they may be powered by solar energy.

Chapter 9 tables

Chapter 9 tables are referenced throughout the NEC. We will take a look at a few tables that relate to PV system design that will let us know about voltage drop and how many conductors we can physically fit in conduit.

Chapter 9 Table 1 Percent of Cross Section of Conduit and Tubing for Conductors and Cables is used to determine how much of the cross sectional area of a conduit can be used for wire and how much extra space needs to be left over for air, cooling and pulling wires. This table is really about geometry and how many circles can fit in a circle. There are three categories in this table. One conductor in a conduit, which is unusual, can take up 53% of the space inside of the conduit. Two conductors in a conduit, which is also not common can only take up 31% of the cross sectional area in the conduit. When we have three or more conductors in conduit, which is usually the case, we can take up 40% of the cross sectional area of the conduit.

Chapter 9 Table 4 Dimensions and Percent Area of Conduit and Tubing is used to determine the interior cross sectional area of conduit and can be used with Chapter 9 Table 1. This table covers many types of conduit and is many pages long. Additionally, this table gives us the data for the percentages of cross sectional area that are required by Chapter 9 Table 1, so we do not need to even look at Chapter 9 Table 1 when we are using Chapter 9 Table 4.

For example, the common conduit used for residential PV projects is 3/4-inch EMT and we can use Chapter 9 Table 4 to see that 40% of the interior cross sectional area of **3/4-inch EMT is 0.213 square** inches. This is how much space we can use without wires and we will see how much space the wires take up next.

Chapter 9 Table 5 Dimensions of Insulated Conductors and Fixture Wires can be used to determine the cross sectional area dimensions of wires. Like Chapter 9 Table 4, this table is many pages long. This table is also used with Chapter 9 Table 4.

An example of the dimensions of a commonly used wire in the PV industry would be using this table to determine the cross sectional dimensions of 10 AWG THWN-2. We go down the left column until we see THHN, THWN, THWN-2, which all have the same dimensions and we can see that **10 AWG THWN-2 has a cross sectional area of 0.0211 square inches.**

If we use our examples from Chapter 9 Tables 4 and 5, we can divide the cross sectional area of 3/4-inch EMT, which is 0.213 square inches by the cross sectional area of 10 AWG THWN-2, which is 0.0211 square inches and we get:

0.0213 / 0.0211 = 10.1 conductors

We have to round down to 10 conductors in this example to fit in this conduit. We will see a little later in this chapter that we can use Informative Annex C to cut down on math.

Rounding up for number of conductors in conduit

Chapter 9 Note 7 tells us that we can **round up if we are 0.8 over a whole number** when calculating the number of conductors that will physically fit inside a conduit.

People in the field do not like using the maximum number of conductors that can fit in a conduit, because a tight fit can be very difficult to work with. It is recommended to have extra space.

Chapter 9 Table 8 Conductor Properties is used often to calculate voltage drop, but there are also other convenient properties of this table. In our discussion of wire sizing in chapter 12 of this book, we will use Chapter 9 Table 8 for calculating voltage drop, using the resistance values in ohm per kilofoot (kFT) in order to determine the resistance of a wire.

For instance, we can see that an uncoated copper stranded 10 AWG wire has a resistance of 1.24 ohms per kFT. This means that 1000 feet of 10 AWG wire has a resistance of 1.24 ohms, so 500 feet would have half of that. One thing in this table that often confuses people is the uncoated vs. coated column in this table. The **coated wire is usually a copper wire that was dipped and has a tin-plating on it to help with** corrosion. We most often do not use coated wire and this **coating has nothing to do with the wire having insulation,** as is often confused.

Chapter 9 Table 8 also has the metric cross-sectional area of the wire. **In most of the world, square mm cross-sectional area is used to designate a wire's dimensions** rather than the AWG system. Here we can use this table to convert.

We can also see wire dimensions in circular mils. **A circular mil is the cross sectional area of a circle with a diameter of 1/1000 of an inch** (very small). Often times **larger wires are measured in thousands of circular mils (kcmil).**

Chapter 9 Table 9 Alternating-Current Resistance and Reactance for 600-Volt Cables, 3-Phase, 60Hz, 75°C – Three Single Conductors in Conduit is used to calculate voltage drop for ac voltage drop for larger conductors. With larger conductors, there is a greater difference between ac and dc with regards to voltage drop.

Skin effect is when the alternating current has a tendency to ride on the outer "skin" of the wire with larger wires, thus making ac less efficient than dc with larger wires. This is one reason that larger wires are run in parallel, such as when you see multiples of three wires running along huge utility power poles.

There are also more variables with these calculations, such as the type of conduit, the power factor and copper vs. aluminum wire.

With wires smaller than 250 kcmil, the difference between using Chapter 9 Table 8 and Chapter 9 Table 9 is usually less than 5% more voltage drop (not voltage drop percentage).

Table 10 deals with the number of strands in typical cables. The most common conductors use Class B stranding. Standard terminals are designed for Class B stranded conductors. For higher strand counts, as is required for tracking systems in 690.31(E), special connectors are required to properly terminate the conductors. Most PV modules have Class C or similar stranded PV wire for the module leads.

Informative annexes

Further in the back of the NEC are Informative Annexes and Informative **Annex C Conduit and Tubing Fill Tables for Conductors and Fixture Wires of the Same Size** (long name) is especially useful for determining how many conductors fit in a conduit, in many cases avoiding the math involved with using the tables in Chapter 9.

If we wanted to arrive at fitting ten 10 AWG THWN-2 conductors in a 3/4-inch EMT conduit, we can look for THWN-2 in the left column of Informative Annex C and a few pages in we see THWN-2, we match 10 AWG with 3/4-inch EMT and can see that 10 conductors fit. Informative annex C does not work for conductors of different sizes within the same conduit.

Informative Annexes are not part of the requirements of the Code, but contain helpful shortcuts.

Index

Last but not least in the NEC is the Index, which contains everything from ac to zones. Become familiar with the index and use it often, especially if you are planning on taking an open book NEC exam, or need to fall asleep. The smartest people know how to use the index.

Since we are talking about the index here, "index" shall be in the index of this book.

We have now covered PV and the NEC. The next and last chapter will use the NEC by working out wire sizing examples.

12 PV wire sizing examples

Wire sizing has been so complicated that many experts disagree on how to correctly size a wire and many books have conflicting methods for how to properly size a wire.

In sizing a wire there are many different checks that should be done. Some of the checks seem so obvious that they are usually skipped and other checks are sometimes just given a brief statement, such as "then check that the overcurrent device satisfies Article 240 Overcurrent Protection."

We will give a few examples of wire sizing and then let you practice on your own. Practice makes perfect. Someone who sizes PV wires every day will often skip most checks, since they know from experience which check will determine the wire size in their particular situation.

Wire sizing example 1

Inverter output circuit wire sizing

Given the following information:
  Inverter continuous rated output current = 10A

- If it were on a house, then 240V × 10A = 2.4kW inverter

Number of current-carrying conductors in conduit 2

- Do not include ground or balanced neutral

ASHRAE 2% high temperature from www.solarabcs.org = 40°C
  Distance above roof conduit in sunlight = 1 inch
  Terminal temperature limits = 75°C

190   PV wire sizing examples

- Terminals are what we attach ends of wires to.

Wire type to be used = THWN-2

- 90°C rated wire
    - We can see this in Table 310.15(B)(16)
    - -2 or at end of wire designation means 90°C rated

Discussion
Defining current for this exercise:

690.8(A)(3)

Maximum circuit current = continuous rated output current = 10A = Imax

690.8(B)(1)

Required ampacity for continuous current = Imax × 1.25 = 12.5A = Icont

This example is a 90°C rated wire and a 75°C rated terminal.

We are going to break this down into 10 steps (at least it's not 12 steps although you might need to recover with a 12-step program after we are done). Some of the steps will seem useless in most cases, but it is possible to have a 75°C rated wire with 90°C rated terminals, although we have never seen it happen. We could make fewer steps and then find an unusual exception where the four-step process does not work.

Legend:

COU = Conditions of Use (adjustment or derating factors)
Ampacity = Conductor's ability to carry current
Imax = Maximum Circuit Current 690.8(A)
Icont = required ampacity for continuous current = Imax × 1.25
OCPD = Overcurrent protection device

PV wire sizing examples 191

10 steps

1. Round up Icont to fuse size
2. Pick conductor size (maybe from 240.4D)
3. 75°C ampacity
4. 75°C ampacity ≥ Icont good!
5. 75°C ampacity >= OCPD good!
6. 90°C ampacity
7. 90°C ampacity >= OCPD good!
8. 90°C ampacity × COU deratings = COU derated wire
9. COU derated wire >= Imax good!
10. COU derated wire round up to OCPD >= OCPD from step 1 good!

Working the 10 steps with our example 1

1. Round up Icont to fuse size

   - 12.5A rounds up to 15A
   - 240.6

2. Pick conductor size

   - Educated guess or from 240.4(D)
   - 14 AWG copper is smallest wire from 240.4(D).

3. 75°C ampacity (75°C terminals)

   - 14 AWG = 20A
   - Table 310.15(B)(16) conduit or (B)(17) if free air

4. 75°C Ampacity >= Icont good!

   - 20A >= 12.5A good!
   - If not good, increase the conductor size here

5. 75°C Ampacity >= OCPD good!

   - 20A > 15A good!
   - If not good, increase the conductor size

6. 90°C ampacity (90°C rated wire)

   - 14 AWG = 25A
   - Table 310.15(B)(16) conduit or (B)(17) if free air

7. 90°C ampacity >= OCPD good!

   - 25A >= 15A good
   - If not good, increase the conductor size here

PV wire sizing examples

8. 90°C ampacity × COU deratings = COU derated wire
   - COU = Conditions of Use
   - COU deratings from 690.31(A) and 310.15(B)(3)(a)
   - 690.31(A) 40°C for 90°C rated wire = 0.91 derating
   - 310.15(B)(3)(a) no derating for two current-carrying conductors
     - Do not count balanced neutral and ground
   - 25A wire × 0.91 = 23A rounded to nearest whole number
9. COU derated wire > Imax good!
   - 23A > 10A
10. COU derated wire round up to OCPD >= OCPD from step 1 good!
    - 23A wire rounds up to 25A
    - 25A >= 15A

Conclusion: 14 AWG wire satisfies the requirements of the code in this case, however, most people would use a larger wire due to voltage drop. Ten years ago, everyone used a 10 AWG wire, still many people use a 10 AWG wire, but with less expensive PV 12 AWG wire is more common and 14 AWG wire does not violate the NEC.

> Step 10, Rounding up wire to common overcurrent device
>
> One of the most difficult concepts for people to get is that in step 10, we can take the ampacity of the wire and round up to the next common overcurrent protection device size. This is not an actual overcurrent protection device; this is just saying that a wire that is derated to 23A can actually handle 25A because there is extra ampacity built into 310.15(B)(16) and a 23A wire in reality can handle 25A.

Wire sizing for voltage drop is a good idea, but is never a Code issue with the NEC. We will do voltage drop calculations later in this chapter, after we focus on Code compliant wire sizing.

Example 2

PV source circuit wire sizing

Sizing a PV source circuit given the following information:

PV wire sizing examples

Isc = 8A
Number of PV source circuits in a conduit = 20
ASHRAE 2% high temperature from www.solarabcs.org = 40°C
Distance above roof conduit in sunlight = 1 inch
Terminal temperature limits = 75°C
Wire type to be used = THWN-2

Discussion

Defining current:

690.8(A)(1)

Maximum circuit current = Isc × 1.25 = 8A × 1.25 = 10A = Imax
(Imax is different from and not to be confused with Imp)
Required ampacity for continuous current = Imax × 1.25 = 12.5A = Icont
(Icont = Isc × 1.25 × 1.25 = Isc × 1.56)
THWN-2 = 90°C rated wire and we are using 75°C terminals as mentioned.
20 PV source circuits = 40 current-carrying conductors

Working the 10 steps with our example 2

1. Round up Icont to fuse size
   - Icont = 12.5A rounds up to 15A fuse as per 240.6

2. Pick conductor size
   - 15A fuse requires at least 14 AWG copper as per 240.4(D)

3. 75°C ampacity (75°C terminals)
   - 75°C 14 AWG = 20A as per Table 310.15(B)(16)

4. 75°C Ampacity >= Icont good!
   - 20A >= 12.5A good!

5. 75°C Ampacity >= OCPD good!
   - 20A > 15A good!

6. 90°C ampacity (90°C rated wire)
   - 14 AWG = 25A per Table 310.15(B)(16)

7. 90°C ampacity >= OCPD good!
   - 25A >= 15A good

194 PV wire sizing examples

8. 90°C ampacity × COU deratings = COU derated wire
   - 690.31(A) 40°C for 90°C rated wire = 0.91 derating
   - 310.15(B)(3)(a) for 40 conductors in conduit = 40% = 0.4
   - 25A wire × 0.91 × 0.4 = 9A rounded to nearest whole number

9. COU derated wire > Imax good!
   - 9A is *not* > 10A so go back to use next larger wire 12 AWG
   - **12 AWG = 30A** per Table 310.15(B)(16)
   - 30A × 0.91 × 0.4 = 11A rounded to nearest whole number
   - **11A > 10A** (notice we are not using Icont here)

10. COU derated wire round up to OCPD >= OCPD from step 1 good!
    - **11A wire rounds up to 15A** as per 240.6
    - 15A >= 15A

Conclusion: **12 AWG satisfies the requirements of the Code here**. It is interesting to note that the conditions of use rated wire is 11A and we can round that up to 15A and have an 11A wire protected by a 15A overcurrent protection device! If you go to Europe, you will see that their wires can carry more current for the same size wire than AWG wires can. We have a buffer of protection built into our wires that will let us deny common sense and round up a wire's ability to carry current.

Would we use a 12 AWG wire here in reality? I think I would use a 10 AWG wire, just to be safe and simple. We do not want to push our luck here with what we have learned in chapter 12.

Voltage drop

When it comes down to voltage drop, what we really want to know is how much money our wire will save for us if we invest more money in the wire. There will be complex calculations, which would have to include tilt, azimuth, soiling, PV to inverter ratio and weather. In order to perform those calculations, it is recommended to use complex software and perhaps to hire a team of engineers.

For the purposes of this book, we will use the maximum output current of the inverter, which is being very conservative, since most if not all of the energy generated from a PV system is going to be less than the maximum output current. For PV source and PV output circuits, we will use the current at maximum power (Imp), which is considerably less than the currents we used to calculate Code compliant wire sizes and is more than we will often see on a PV source circuit.

Some designers will use 80% of these numbers as a rule of thumb, since most of our energy is made when it is not a cold, windy, bright summer noon (optimal PV conditions). We will use Imp and inverter maximum output current for this book, which is conservative and leads to less energy loss over the year than voltage drop percentage in the calculation.

If you are performing voltage drop calculations for a job that you won a bid on or are bidding on, you should carefully read the requirements of the request for proposal.

We will use a simple calculation to arrive at an AWG wire size given the following information:

Voltage = 240V
Current = 16A
Voltage Drop Percentage = 2%
Distance from inverter to interconnection = 200 feet

Here is the formula that can be used with Chapter 9 Table 8 of the NEC

Ohms/kFT = (5 × % × V)/(I × L)
Ohms/kFT will give us an AWG wire size in Chapter 9 Table 8
5 is a constant derived from (1000FT/kFt)/100%/2 wires in a circuit)
% is the percentage, so we use 2 (not 0.02) for 2%
V is the operating voltage, which is 240V at your house
I is the current of the inverter in this case, which is 16A for a 3.8kW inverter
L is the 1 way distance in feet which is 200 FT

We will plug it in to the equation:

Ohms/kFT = (5 × % × V)/(I × L)
Ohms/kFT = (5 × 2% × 240V)/(16A × 200FT) = 2400/3200 = **0.75 ohms/kFT**

If we **look up 0.75 ohms/kFT in Chapter 9 Table 8** we see that an uncoated 6 AWG copper wire will have a resistance of 0.491 ohms/kFT and a smaller 8 AWG stranded copper wire will have a resistance of 0.778 ohms/kFT.

Since voltage drop is not a Code issue here, you can choose to round up or down from a 6 AWG or an 8 AWG wire.

This calculation will work for ac and dc wires because the values in Table 9 are essentially the same for ac circuit running at unity power factor. If you are using a large wire for ac and running the circuits at a

196 PV wire sizing examples

power factor of 0.85 (may be required occasionally by utilities for grid support), then the values in Table 9 differ from those in Table 8. It's best to get an engineer involved for larger systems as these calculations can get complicated.

In order to use these calculations for 3-phase power, just remember that there is a benefit to using 3-phase that is proportional to the square root of 3 (about 1.73). If we divide the square root of 3 by 2 we get 0.886, so we will have 88.6% of the resistance with 3-phase wires or we can multiply our ohms/kFT answer by 0.866. In the example we used, instead of 0.778 ohms/kFT, we could use a wire that is 0.778 × 0.866 = 0.67 ohms per kFT for 240V 3-phase.

The reason we divide the square root of 3 by 2 is because with 3-phase our currents are not directly opposing each other (square root of 3) and we are converting from a calculation that is from single phase power where we have to double the one-way distance of our wire to calculate the resistance of a circuit.

A circuit is a circle and if you are going to have your inverter 200 feet from the interconnection, you need to run electrons through 400 feet of wire and will have 400 feet = 0.4 kFT of resistance. With 3-phase, you will need to have current on three wires, but it will be less current, since the currents are 120 degrees out of phase with each other.

Figure 12.1 Nicola Tesla demonstrates how to truly understand 3-phase in 1899

Some people say that understanding 3-phase power takes more than a lifetime to truly understand, but if Tesla (a crazy genius) could figure out how 3-phase power works all on his own, you can too!

Thank you for reading this book!

Sean and Bill

Index

Page numbers in *italics* indicate a figure on the corresponding page; page numbers in **bold** indicate a table on the corresponding page

ac (alternating current) module: definition 13–14; modules (690.6) 24–25; photovoltaic modules (690.52) marking 125; system *10*
accessible, readily, definition 75
access to boxes (690.34) 102
ac coupled multimode system *13*
ac microgrid 12
additional auxiliary electrodes for array grounding 120–121
adjustable electronic overcurrent protective device 46–47, 54
Adjustable-Trip Circuit Breakers (240.6) 46
adjustment factors, for conductor ampacity 44–46
Advanced Energy Inverters 110
Allowable Ampacities of Insulated Conductors **182**
ambient temperature correction factors 45, 46, **182**
American Wire Gauge (AWG): conductor cables 97; equipment grounding conductor 116, 117
ampacity 42, 97; of neutral conductor (705.95) 164–165
ampere interrupting rating 78–79
amp rating 78–79
ANSI (American National Standards Institute) 131; equipment certification 112; field-applied hazard markings 76–77
AP system 4 module inverter *63*

arc-fault circuit protection 56; exception 56–57
arc-fault mitigation (691.10) 136
array, definition (690.2) 60
array boundary: definition (690.2) 60; inside the 62–64; outside the 61–62
Article 100 definitions 178
Article 110 requirements for general installations 178
Article 200 grounded conductors 178
Article 225 93–94
Article 230 services 178
Article 240 Overcurrent Protection 34–35, 49, 179, 189
Article 250 Grounding and Bonding 2, 179–180
Article 300 wiring methods and materials 181
Article 310 2; conductors for general wiring 181–182
Article 340 94
Article 352 rigid polyvinyl chloride conduit 183
Article 392 cable trays 183
Article 400 Flexible Chords and Cables 96–97
Article 450 Transformers and Transformer Vaults 160
Article 480 Storage Batteries 1, 169–170, 183; disconnecting means 170; live parts 170; scope

169–170; spaces about battery systems 170; ventilation 170

Article 625 Electric Vehicle Charging Stations 175

Article 690 Photovoltaic (PV) Systems 1; access to boxes (690.34) 102; alternating current (ac) modules (section 690.6) 24–25; arc-fault circuit protection (section 690.11) 56–57; circuit sizing and current (section 690.8) 34–49; component interconnections (690.32) 101; connectors (690.33) 101–102; definitions (section 690.2) 13–22, 64; energy storage systems 128; equipment bonding jumpers (690.50) 121–122; equipment grounding and bonding (690.43) 113–115; general requirements (section 690.4) 22–24; grounding electrode system (690.47) 117–121; maximum voltage (section 690.7) 26–33; outline of 4–6, 58–59; overcurrent protection (section 690.9) 49–55; Part VI (of the NEC) marking 123–127; Part VII (of the NEC) connection to other sources 128; Part VIII (of the NEC) energy storage systems 128; point of system grounding connection (690.42) 113; rapid shutdown of PV systems on buildings (section 690.12) 59; scope (section 690.1) 7–13; self-regulated PV charge control 128–129; size of equipment grounding conductors (690.45) 115–116; special equipment 184; stand-alone systems (section 690.10) 56, 168; system grounding (690.41) 103–112; wiring methods (690.31) 89–101

Article 691 Large-Scale Photovoltaic (PV) Electric Power Production Facility 1, 18, 56; arc-fault mitigation 136; conformance of construction to engineered design 135; definitions 132; direct-current operating voltage 135; disconnection of photovoltaic equipment 136; engineered design 135; engineering review 134–135; equipment approval 134–135; fence grounding 136; field labeling 134; information notes 131–132; medium or high voltage connection 133–134; outline of 130–131; overview 136–137; qualified personnel 132–133; restricted access 133; scope 131; special requirements 132–134

Article 702 Optional Standby Systems 175, 185

Article 705 Interconnected Electric Power Production Sources 1; ampacity of neutral conductor 164–165; circuit sizing and current 162–163; definitions 141; directory 142–143; disconnect device 159; disconnecting means, equipment 159; disconnecting means, sources 158; equipment approval 142; ground-fault protection 161; grounding 162; interactive inverters in not readily accessible locations 163–164; interactive system disconnecting means 160; interrupting and short-circuit rating 158; location of overcurrent protection 160–161; loss of 3-phase primary source 162; loss of primary source 161; other articles 142; outline of 138–141; output characteristics 158; overcurrent protection 160, 163; point of connection 143–158; scope 141; system installation 142; unbalanced interconnections 165; *see also* point of connection (705.12)

Article 706 Energy Storage Systems 1, 168, 170–173; definitions 171; disconnecting means/notification 171–172; dwelling units 172; installation of batteries 172; scope 170

Article 710 Stand-Alone Systems 1, 56, 173, 184

Article 712 Direct Current Microgrids 1, 168, 174–176, 184; definitions 174–175; system voltage 175–176

Index

ASHRAE Handbook 27
auxiliary electrodes 119

backfeeding 17, 48; overcurrent protective device preventing 54; shorted PV source circuit 52, 53
batteries: energy storage and 168–169; matching PV to 129
bimodal inverter 11
bipolar arrays 14
bipolar photovoltaic array: definition 14; grounding configuration 106, *107*; systems 100–101
bipolar source and output circuits 33
Bower, Ward 3
breakers over 1000V prices *159*
Brooks, Bill 3, 27
Building Integrated Photovoltaics (BIPV) definition 64
busbars (705.12) 151, 153–157; 100% rule 151, *153*; 120% rule 153–154, *155*; center fed 120% rule 156–157; marking 157; multi-ampacity busbars with engineering supervision 157; sum rule 154, *155*, 156

cable trays, Article 392 183
cable wiring methods 89, 90
circuit current calculation of maximum 35–38
circuit sizing and current (690.8): calculation of maximum circuit current 35–38; conductor ampacity 41–47; dc-to-dc converter source circuit current 40–41; defining currents 34; inverter output circuit current 38–40; outline of 34; overcurrent protection 34–35; photovoltaic output circuits 38; sizing of module interconnection conductors 47–49; stand-alone inverter input circuit current 40; systems with multiple direct-current voltages 47
circuit sizing and current (705.60) 162–163
Code (the) 3, 15, 54, 58, 76, 86, 100, 110, 113, 117, 120, 146, 163, 167, 174–175, 188

combiner 16, 20, 27; ac 154; dc 20, 23, 27, 38, 52, 83
communication systems 185
component interconnections (690.32) 101
conductor ampacity 41–47; adjustable electronic overcurrent protective device 46–47; application of adjustment factors 44–46; overcurrent protection 163
Conductor Ampacity Code References 41
conductors: length of free 90; properties 186–187
conduit and tubing: conductors and cables 185; dimensions and percent area of 185
conformance of construction to engineered design (691.7) 135
connectors (690.32) 101–102; configuration 101; grounding member 102; guarding 102; interruption of circuit 102; type 102
continuous current, definition 35
current-limited: inverter 55; PV system 49–52, 82, 111, 172

Database of State Incentives for Renewables and Efficiency 176
dc combiners 52
dc coupled multimode system *11*
dc-to-dc converter: definition 15; PV and energy storage systems 33
dc-to-dc converter output circuit current 41
dc-to-dc converter output circuits, definition 16
dc-to-dc converter source and output circuits: calculations 32–33; definition 15
dc-to-dc converter source circuit current 40–41
direct current microgrid (712) 168, 174–176
direct-current operating voltage 135
direct-current photovoltaic power source (690.53), marking 125–126
directory (705.10) 142–143
disconnect device (704.22) 159
disconnecting means 10, 24; definition 74; directional current

202 Index

devices 175; equipment (705.21) 159; outline of (section 690.13) 74–75; PV system 65; service 65, 70; sources (705.20) 158; storage batteries 170; types of 86–87; *see also* disconnection of photovoltaic equipment (690.15); photovoltaic system disconnecting means (690.13)

disconnection of photovoltaic equipment (690.15) 80–87; equipment disconnecting means 84–87; finger safe fuse holder 84; interrupting rating 82; isolating device 81, 82–83; load-break rated 81; location 81–82; non-load-break disconnect 81; outline of 80

disconnection of photovoltaic equipment (691.9) 136

diversion charge controller, definition 16

Earth's atmosphere, irradiance outside 37
Edison, Thomas 1
electrical connections, flexible, fine-stranded cables 100
electrical engineer, licensed professional 32, 37–38
electrical metallic tubing (Article 358) 183
electrical production and distribution network, definition 16
Enclosure for Electrical Installations (110.31) 133
energy storage, batteries and 168–169
energy storage system (ESS): definition 171; ESS (Article 690 Part VIII) 128; ESS (Article 706) 128
engineered design (691.6) 135
engineering supervision method, calculating maximum circuit current 37–38
equipment approval (705.6) 142
equipment bonding jumpers (690.50) 121–122
equipment grounding and bonding (690.43) 113–115; with circuit conductors 115; equipment secured to grounded metal supports 114–115; outline of 113; photovoltaic module mounting systems and devices 114; why equipment grounding 114
equipment grounding conductor (EGC): Article 100 definition 118; size of 115–116
Expedited Permit Process 27–28, 45

fastening (705n12) 158
feeder taps (240.21) 179
fence grounding (691.11) 136
field-applied hazard markings (110.21) 76–77
field labeling (691.5) 134
finger safe fuse holder 83, 84
fixture wires, dimensions of 186
flexible, fine-stranded cables 100
flow batteries 173; definition 171
"formerly known as grounded" 17
"formerly known as ungrounded" 17–18
fuel cell systems, overcurrent protection 160
functional grounded PV system: definition 16–18; "formerly known as grounded" inverters 17; "formerly known as ungrounded" inverters 17–18; term 18
functional grounding, term 18
functional ground inverter 106
fuse grounded current-carrying conductors 93
fuse grounded inverters 109
fuses, PV source circuit 51, 52, 53

generating capacity: definition 18; definition (691.2) 132
generating station, definition (691.2) 132
generators, overcurrent protection 160
GFDI (ground fault detection and interruption) 17, 104, 106
ground 118
grounded 2-wire dc system, definition 174
grounded conductor, Article 100 definition 118; identifying 92–93,

Index 203

95; use and identification (200) 178
grounded inverters 85–86, 104
ground fault, Article 100 definition 118
ground fault circuit interrupter (GFCI), Article 100 definition 118
Ground Fault Detection (690.41) 111, 112
Ground Fault Protection (690.41) 111
ground-fault protection (705.32) 161
ground fault protection exception (690.41) 111
grounding and bonding (250) 179–180
grounding electrode, Article 100 definition 118
grounding electrode conductor (GEC), Article 100 definition 118
grounding electrode system (690.47): additional auxiliary electrodes for array grounding 120–121; buildings or structures supporting PV array 118–119; informational note 120; installation 180; outline of 117; Part III of Article 250 119; permitted 180; requiring the following of rules 119–120

hard service chord 96
hybrid system, Article 100 definitions 12; hybrid systems (705.82) 164

identification of power sources (690.56), marking 127
identified = for specific use (no off-label usage) 115; cable ties 94; chords and cables 96; isolating device 82; multiconductor cable 96; PV module equipment 114; PV system circuit conductors 92–93; rapid shutdown 59
IEC (International Electrotechnical Commission) 15, 50, 109
industry standard method 37–38; calculating maximum circuit current 38; calculating maximum voltage 32
informational note 23, 32

informative annexes 187–188
initiation device, rapid shutdown 65–66
insulated conductors, dimensions of 186
interactive inverter output circuit, definition 19, 141
interactive inverters, overcurrent protection 160
interactive source of interconnection (690.54), marking 126
interactive system 10; definition 19
inverter grounding, 2014 vs. 2017 NEC 85–86
inverter output circuit current 38–40; wire sizing 189–192
Isc (short-circuit current) 36, 37, 38
Isolated Faulted Circuits (690.41) 111, 112
isolating switch: definition 81; requiring tool 83

labeled = has a label from the National Recognized Testing Lab (NRTL) 115; equipment 22, 23, 142; PV module 114, 134; rapid shutdown 62; wiring system 93
listed = on a NRTL list 115; arc-fault circuit protection 56; articles 142; cable ties 94; component interconnections 101; connectors 86; directional current devices 175; equipment 22, 66, 114, 122, 142; flexible chords and cables 96; fuses for PV 53; ground fault detection 112; inverters 158, 162, 165, 167; isolating device 82, 83; multiconductor cable 96; overcurrent protective device 42, 54, 144; PV module 10, 11, 114, 134; PV panels 23; PV source circuits 25; rapid shutdown 59, 62; supply side disconnecting means 77; ungrounded inverters 109; wiring system 90, 93
lithium batteries 14
load-break rated 81
load side (705.12) 143–158; 10-foot tap rule for solar 148, 150; bus or conductor ampere

204 Index

rating 144–157; dedicated overcurrent and disconnect 144; feeder ampacity protection 146; feeder overcurrent device *147*, 148; feeders 146, *146*; taps 148, *149*, 150
location of overcurrent protection (705.31) 160–161

marking (110) 178
marking (690 Part VI) 123–127; alternating-current photovoltaic modules (690.52) 125; direct-current photovoltaic power source (690.53) 125–126; identification of power sources (690.56) 127; interactive source of interconnection (690.54) 126; modules (690.51) 124–125; outline of 124; photovoltaic systems connected to energy storage systems (690.55) 126
marking (705.12) 157
maximum circuit current *37*; engineering supervision method for calculating 37–38
maximum voltage (690.7): engineering supervision method 32; informational note 27–32; outline of 26; photovoltaic source and output circuits 27; table method 30–31; voltage temperature calculation method 28–30
microgrid interconnect device (MID), definition 141, 167
microgrid system: alternating current 166; definition 141; system operation 166
microinverter 11
module interconnection conductors, sizing of 47–49
module level shutdown 62–63
modules (690.51), marking 124–125
monopole subarrays 100
MPP (maximum power point) 21, 33
MPPT (maximum power point tracking) 21, 27
multiconductor cable 96
multimodal inverters 11, 12; definition 19, 141

multiple direct-current voltages 47
multiple PV systems (690.4) 24

National Electric Code (NEC) 1–3; 1984 NEC book *5*; 2014 NEC PV power source *8*; 2014 *vs.* 2017 inverter grounding 85–86; 2017 NEC ac coupled multimode system *13*; 2017 NEC PV power source *9*; index 188; informative annexes 187–188; tables in 185–187
National Electric Safety Code (NESC) 130, 131
National Fire Protection Association (NFPA) 3, 133; NFPA 70E Standards for Electrical Safety in the Workplace 133, 172
Nationally Recognized Testing Lab (NRTL) 134
negative connector 126–127
neutral conductor (705.95), ampacity of 164–165
Niagara Falls: first power plant 3n1; Niagara Falls power plant (1895) *2*
nominal voltage definition 174
non-isolated array 109
non-isolated inverter 106, 108, *108*
non-load-break disconnect 81

Overcurrent Device Ratings (690.9) 115
overcurrent protection (240) 179
overcurrent protection (690.9): circuits and equipment 49–50; exceptions 51–52; informational note 52, 54, 55; outline of 49; overcurrent device ratings 54; photovoltaic source and output circuits 54–55; power transformers 55; strings 50
overcurrent protection (705.30) 160
overcurrent protection (705.31): exception method 161; location of 160–161; method 160
overcurrent protection (705.65) 163

Part VI marking (690) 123–127; alternating-current photovoltaic modules (690.52) 125; direct-current photovoltaic power

Index 205

source (690.53) 125–126; identification of power sources (690.56) 127; interactive source of interconnection (690.54) 126; modules (690.51) 124–125; photovoltaic systems connected to energy storage systems (690.55) 126
photovoltaic (PV) arrays, wiring systems for 90
photovoltaic (PV) module parallel-connected circuits 47–49
photovoltaic (PV) output circuits 38, 91–92; definition 20–21
photovoltaic (PV) power source: definition 21; marking and labeling 99–100; NEC (2014) 8; NEC (2017) 9
photovoltaic (PV) source circuit 91–92: 3-phase power 196–197; backfeeding a short on PV source circuit 53; definition 20; engineering supervision method 37–38; fuses 51, 52, 53; short-circuit current method 36–37; voltage drop 194–196; wire sizing 192–197
photovoltaic (PV) source circuit (string) voltage: calculation method 28–30; engineering supervision method 32; methods for determining 28–32; table method 31–32
photovoltaic (PV) system disconnecting means (690.13): field-applied hazard markings (110.21) 76–77; location 75–76; marking 76–77; maximum number of disconnects 78; outline of 74–75; ratings 78–79; suitable for use 77; type of disconnect 79–80; *see also* disconnection of photovoltaic equipment (690.15)
photovoltaic (PV) systems 1–3: connected to energy storage systems marking 126; direct current circuit 21, 97–100; identifying circuit conductors 92–93; PV systems (690.4) 22; self-regulated charge control 128–129; special occupancies 183–184

photovoltaic (PV) wire sizing: inverter output circuit wire sizing 189–192; PV source circuit wire sizing 192–197; wiring methods 94–95
point of connection (705.12): busbars 151, 153–157; bus or conductor ampere rating 144–157; dedicated overcurrent and disconnect 144; fastening 158; feeders 146–148; load side 143–158; marking 157; suitable for backfeed 157; supply side 143; taps 148–151; *see also* Article 705 Interconnected Electric Power Production Sources
point of system grounding connection (690.42) 113
polyvinyl chloride (PVC), rigid conduit 183
portable power cable 96
positive connector 126–127
power optimizers 15
power production equipment, definition 141
power sources, identification of 127
power transformers, overcurrent protection for 55
primary power source: connection 166; reconnection to 166

qualified personnel (690.4) 23
qualified personnel (691.4) 132–133

raceway wiring methods 89, 90
rapid shutdown (section 690.12) 59, 136; 2014 NEC switch label differences 72–73; controlled conductors 60; controlled limits 60–64; equipment 66–67; exception 60–67; informational note 65–66; initiation device 65–66; initiation switch 66; labeling (from 690.56) 67–72; marking buildings with 127; methods for initiation of 65; outline of 2014 NEC (690.12) 59; outline of 2017 NEC (690.12) 58–59; TIA (Tentative Interim Amendment) 59; *see also* rapid shutdown labeling (690.56)

rapid shutdown labeling (690.56): 2014 NEC switch label differences 72–73; buildings with more than one type 71–72; conductors leaving array level shutdown 69–70; outline of 67; rapid shutdown type 67–70; reduced array shock hazard sign 68; switch 72
readily accessible, definition 75
reference-grounded system, definition 174
resistively grounded, definition 175
restricted access (691.4) 133
Restricted Access Adjustable-Trip Circuit Breakers (240.6) 46
rigid polyvinyl chloride conduit (Article 352) 183

Sandia National Laboratory 32
self-consumption 12
self-regulated V charge control (690.72) 128–129
Service-Entrance Cable, USE–2 90
single conductor cable 93–95
single-phase inverters 165
size of alternating-current grounding electrode conductor 119
size of direct-current grounding electrode conductor 120
size of equipment grounding conductors (690.45) 115–116
skin effect 187
small conductor cables 97
small conductor rule 179
smart electronics 35
Solar America Board of Codes and Standards 27
solar cells, in PV power sources *8, 9, 9*
solar panel, slang term 19
solar panels *vs.* solar modules 19–20
solar photovoltaic systems: equipped with rapid shutdown 68–73; overcurrent protection 160
solidly grounded, Article 100 definition 118
solidly grounded photovoltaic array 110, *111*
solidly grounded systems 92–93

stand-alone inverters 39; input circuit current 40; partial datasheet from outback *39*
stand-alone systems *14*; Article 710 173; definition 22; marking facilities with 127; wiring of 56
storage articles: Article 480 storage batteries 169–170; Article 706 energy storage systems 170–173; Article 710 stand-alone systems 173; Article 712 direct current microgrids 174–176; batteries and energy storage 168–169
strings 16, 20, 48, *50*
string theory 15–16
supply side connection 78, 143; location of overcurrent protection (705.31) 160–161
supply side disconnecting means 77
switch, isolating, definition 81
system grounding (690.41) 103–104; 2-wire PV arrays with one functional grounded conductor 104, *105*, 106; arrays not isolated from grounded inverter output circuit 106, 108–109; bipolar PV arrays with functional ground center tap 106, *107*; equipment certification 112; ground fault detection 112; ground fault protection exception 111; isolating faulted circuits 112; outline of 103–104; PV configurations 104–111; PV systems using other approved methods 110–111; solidly grounded PV arrays 110, *111*; ungrounded PV arrays 109, *110*

taps (704.12) 148, *149*; 10-foot tap rule for solar 148, *150*; 25-foot tap rule for solar 151, *152*; solar tap rules *150*
terminal temperature 42–43, 44–45, 189, 193
Terminal Temperature NEC Reference 43
Tesla, Nikola 1, *196*, 197
three phase inverters 165

TIA (Tentative Interim Amendment) 59
transformers, overcurrent protection 160

unbalanced interconnections (705.100) 165
Underwriter's Laboratories (UL) 112; UL 1741 22–23, 112, 158, 161, 165, 167; UL 2703 standard 114, 122
ungrounded inverters 85, 106
ungrounded PV arrays: definition of 109; grounding configuration 109, *110*
USE-2 cable: Service-Entrance Cable 90; wiring methods 94–95
utility-interactive power systems employing energy storage 164

wire sizing 189; inverter output circuit 189–192; PV source circuit 192–197
wiring methods (690.31) 89–101; bipolar photovoltaic systems 100–101; correction factors for temperatures over 30°C 91, **91**; flexible, fine-stranded cables 100; flexible chords and cables connected to tracking PV arrays 96–97; identification and grouping 91–93; multiconductor cable 96; outline of methods permitted 89; photovoltaic system direct current circuits 97–100; single conductor cable 93–95; small conductor cables 97; USE-2 and PV wire 94–95; wiring systems 89–91; *see also* component interconnections (690.32)

PGMO 06/26/2018